PRAISE FO

Have just reread your Salish Sea piece and I love it ... it is perfect and beautifully written.

— DR. JANE GOODALL, internationally renowned primate researcher and conservationist

Carbon Play *is a sweeping chronicle of travels, cultures, science and technology unlike any other book on the climate crisis. Roberts Falls's extensive explorations have taken him to the four corners of the earth, and from every different vantage point comes a useful nugget of information on oceans, forests, climate and the technological breakthroughs that make a post-carbon world within reach.*

— ELIZABETH MAY, OC, MP for Saanich–Gulf Islands, Leader of the Green Party of Canada

Carbon Play *is an important, timely and very readable contribution to the dialogue around carbon markets and climate change. The book does not profess to be a comprehensive review of the literature relating to carbon markets and that's to its advantage. Rather, the author successfully meshes science with experiences and accounts from his own personal journey. Such an approach proves to be very successful in making this key and complex issue more accessible to a new and broader audience.*

— MARK ANGELO, founder of World Rivers Day

A delightful personal journey through the carbon landscape. Refreshingly easy to read though dealing with a monumental global issue.

— DR. JOHN WIEBE, founder and chairman of The Globe Foundation

Robert Falls has done well laying out his life as a veteran carbon explorer. I like the way he has combined science (explained in easy to understand language), autobiography (his life has not been boring), and some hopeful and practical solutions for mitigating, adapting and innovating towards a safer future. People like Robert, rather than the climate change deniers or hysterics, will help us limit global warming to 1.5 degrees Celsius.

— MIKE HARCOURT, former mayor of Vancouver, former premier of British Columbia

CARBON PLAY

CARBON PLAY

The Candid Observations of a Carbon Pioneer

ROBERT WILLIAM FALLS

RMB | Rocky Mountain Books Ltd.
rmbooks.com
@rmbooks
facebook.com/rmbooks

Cataloguing data available from Library and Archives Canada
ISBN 9781771602181 (softcover)
ISBN 9781771602198 (electronic)

Cover design by Chyla Cardinal

Printed and bound in Canada by Friesens

Distributed in Canada by Heritage Group Distribution and in the U.S. by Publishers Group West

For information on purchasing bulk quantities of this book, or to obtain media excerpts or invite the author to speak at an event, please visit rmbooks.com and select the "Contact Us" tab.

We acknowledge the financial support of the Government of Canada through the Canada Book Fund and the Canada Council for the Arts, and of the province of British Columbia through the British Columbia Arts Council and the Book Publishing Tax Credit.

Canada

Disclaimer

The views expressed in this book are those of the author and do not necessarily reflect those of the publishing company, its staff or its affiliates.

This book represents some highlights of one carbon pioneer's career path, spanning the years 1978 to 2017.

The book is largely anecdotal. It is not intended to provide a comprehensive review of carbon markets but rather a collection of stories, accounts and insights surrounding one man's journey.

To learn in detail about current carbon markets, and the status of international carbon and climate policies, readers are directed to an ever-growing body of literature readily available from bookstores and online sources.

This book I am dedicating to Linda Gwennyth Kabush Dickinson, who as a young librarian purveyed me the tools and the impetus to write.

Beyond providing me my first journal, complete with a "karma repair kit," a haiku, and some of her own prose, Linda connected me with such literary upstarts as Leonard Cohen, Tom Wolfe, Tom Robbins, Richard Bach and Annie Dillard.

The early '70s were positively alive, and these creative minds, along with the extraordinary illuminations flowing through them, were a big part of the reason why.

Sadly, tragically, Linda passed away on January 9, 2017, a few days before I set off for Thailand to finish the book. She had read my work, all but the final chapter, and I know from our last discussion that she was very happy about it. Even though it had arrived very late, while she felt her own life fleeting, Linda knew her vision would be manifest.

Linda opened the door to brilliant minds, challenge, exuberance and joy.

I am forever grateful.

For the Life of this Planet

The way the red sun surrenders
its wholeness to curving ocean
bit by bit. The way curving ocean
gives birth to the birth of stars
in the growing darkness
wearing everything in its path
to cosmic smoothness.
The impulse of stones rolling
towards their own roundness.
The unexpected comets of flying fish.
And, Forest, Great-Breathing-Spirit,
rooting to the very end
for the Life of this planet.

—Grace Nichols

Table of Contents

Acknowledgements

I first wish to acknowledge Connie Spiers for a range of efforts, from her gentle encouragement to the relentless application of her red pen. Until Connie became involved, the book was a future project – she made sure it happened.

Tianyi Kou took the time from her graduating year at UBC to secure the references of an unfamiliar field, and I thank her for doing so with a constant smile.

Many other people participated in *Carbon Play*, some knowingly, most not. I cannot list them all, but in order of their appearing on my path I wish to thank:

Linda Kabush, librarian extraordinaire, for introducing me to the literature of Man and Nature;

Professor Emeritus Les Lavkulich, Resource Management Science, for revealing to me a viable path forward and challenging me to navigate it;

Professor Iain E.P. Taylor, Curator of UBC's Botanical Garden, for being my first and final biology teacher;

Professor (the late) Glenn Rouse for bringing colour, humanity and light into science;

Dr. (the late) Alex Peden, Curator of Marine Biology at the Royal British Columbia Museum in Victoria, for giving me my first real job as a biologist;

Gordon and Ann Mohs, anthropologists, for sharing their Babine Lake island camp and for some magic moments that continue to mystify;

Debbie Brill, high jumper extraordinaire, for her unmitigated joyfulness, free thinking and athletic discipline;

Jimmie Wright, photographer, for his fun, his appreciation of all things wild, his eloquence and his artistic skills;

Dr. Freeman Dyson, Professor Emeritus at the Institute for Advanced Studies, Princeton, for revealing the irreversible implications of releasing the technological genie from its bottle and for taking the time to confirm and illuminate the Universe's connectedness, that I had stumbled upon.

I also wish to thank my Canadian energy industry colleagues who helped translate the ideal of sustainability into real actions and investments: Bruce Sider, Art Willms, Ken Berry, Wayne Soper, Doug Halverson, Bob Bell, Bill Burton, Eric Mohun, Ed Lee, Suzanne Demitor, Deborah Bisson, Lorna Seppala, Gord McKenzie, Marion Johansen, Timo Makinen, Brian Jantzi, Dawn Farrell, Aldyen Donnelly, John Woodruffe and Marika Hare.

I am honoured to thank (the late) Sir Arthur C. Clarke, the brilliant futurist author, and the eloquent Brian Anderson, Shell Chairman Southeast Asia, for their willingness to explore the geopolitical implications of a low-carbon world 15 years before it was in vogue.

Dr. Patrick Moore and Russ George I appreciate greatly for having the *cojones* to be renegades. Mike Harcourt, former BC premier and mayor of Vancouver, was among the first politicians to embrace sustainability, while bringing the war in BC's forests to a much-needed end.

Professor David Karl, at SOOEST and the University of Hawaii at Mānoa, presented an extraordinary opportunity for me to participate in his tenacious and comprehensive pursuit of ocean science.

I wish to thank ERA co-founder Bart Simmons for sharing his intelligence and his unbounded energy and resolve to always find the way forward, as well as Cornelia Rindt, John Kendall, Joseph Pallant, Erin Kendall, Alex Langer, David Rokoss, Ron Behr and the rest of the ERA people who played their respective roles in inventing an industry;

John Bell and Dr. Ron Dembo invested real dollars into the ERA dream very early in the game and allowed us to prove out our business model.

Graham Harris and Warren Carr expanded our horizons by introducing the carbon play to public markets and to a few world-class musicians, politicians and celebrities who wished to have their carbon footprints addressed with some True North, home-grown, carbon-gobbling ecosystem restorations.

Jeff Horowitz, founder of Avoided Deforestation Partners, and Nobel Laureates (the late) Dr. Wangari Maathai and Al Gore ignited an invaluable reconsideration of the role of forests in climate mitigation and illuminated a "colossal blunder" in international policy that was finally recognized by a host of conservation organizations around the world.

Helen Robinson and her team created the world's first functional international carbon registry, and we were honoured to be among the first to participate in an international trading platform that would ultimately facilitate trillions of dollars in transactions. Bravo.

Dr. Jane Goodall has demonstrated endless patience and perfect kindness with creatures, and her invitation to report some good news about our local restored marine ecosystems for International Oceans Day 2016 revived my resolve to complete the work.

To Bonita Sauder, Karen Fowlie, Agent McPhee and the gang at Hugo's, whose live, homegrown music celebrates life on the Salish Sea, and to the committed stewards of Howe Sound, who protect our gifts of Nature with the vigilance of a mother grizzly bear, many thanks.

And amid the chaos of my carbon career, Corinne and Sebastian and their mother, Diane, gave me a genuine and rich experience of fatherhood that had almost eluded me. Thank you.

And finally, to my sister Cheryl and her family, and my cousins near and far, who never wavered in their support of my serpentine path, I am very grateful.

From Monks to Scientists

1977–1979

CHIANG MAI

As I began to awaken, the sounds of chanting seemed to grow louder and closer. It was the middle of the night and I was alone in a second storey room at a guest house in northern Thailand. I was clearly in a state of delirium – badly dehydrated, running a high temperature and having zero desire or capacity to eat. As the chanting grew louder, I found myself wondering if the sounds I was hearing were originating from something real in the lane below or from the lucid dreaming that accompanied my growing delirium.

Delirium has a way of mixing up reality and the dream world – sometimes it's hard to draw the line between them.

I concluded the chanting was likely real and guessed it was coming from a procession of Buddhist monks moving through the dark streets and alleys in the middle of the night. I either saw, or imagined, flickering lights – candles, likely – dancing among the ephemeral shapes that moved, casting elongated shadows against the walls of dwellings bordering the narrow lane below me.

It was late December of 1977. My original plans had been no more complicated, or naïve, than taking off for six months to travel solo through Thailand and Sri Lanka, facilitated by a backpack, a few clothes and a wad of American Express Traveler's Cheques – the international currency of the time. That was the extent of my plans and preparations. But such an uncharted adventure was likely not an unusual thing for a young man chasing a bigger world view, adventure and perhaps a refreshed raison d'être.

I had arrived in the town of Chiang Mai following a surreal bus ride originating in Bangkok that provided passengers with white tablecloths and American food and played non-stop the elevator Christmas music I had looked forward to escaping from. On arrival, I quickly found a guest house and started wandering around the town and countryside. Other than tepid waterways virtually vibrating with fish life, a visit to a silver jewellery factory and a fruit and yogourt smoothie bar, there is little I can I recall of these daytime forays.

The few evenings of "grace" that I had in Chiang Mai in a state of wellness were spent in bars with some new travelling buddies and local Thai partiers. What was memorable from that experience was sipping the almost colourless "Mekong Whiskey." Being made primarily from sugar cane and molasses, and only a dash of rice, it was closer to being rum than a true rice whiskey. And of course, having been "aged on the truck," it was perhaps not quite as smooth as one might hope for. But it got the job done.

A CHANGE OF COURSE

On the third or fourth day in Chiang Mai, I had treated myself to breakfast – a tropical fruit smoothie. It was and is the best smoothie I have ever tasted. I had another. Late that afternoon my intestinal tract started burbling and gurgling, and I was hit with an intense bout of gastroenteritis. It was during the following night and early morning that the monks were manifest.

Days passed, and I was still very sick with no sign of improvement. Distilled from my muddled thinking came the notion of throwing out what limited plans I had and departing *tout de suite* to a temperate destination – and then figuring things out from there.

Once I was strong enough to get up and walk around, I checked out of the guest house and bused back to Bangkok. There I took a room in a downtown hotel and began to make travel arrangements out of Asia, post-haste. Once again I was awakened in the middle of the night to the surreal: in this case there was no sound, but rather the silent silhouette of a man stealthfully entering my room through a window. When he saw I had awakened, he muttered something, turned around, climbed out through the window and quickly scuttled back from whence he came. I was now on full alert and was on a mission to "get out of Dodge."

SOUTH BY SOUTHEAST

I was so anxious to leave that I bought the first option I saw – an Aeroflot flight from Bangkok to London via Moscow. At C$264, the price was right. Thankfully, the possibility of being delayed in a

Moscow airport during the middle of winter was recognized as insanity and was appropriately tossed. I traded my ticket for a trip to New Zealand that would entail travel by rail, road, water and air, with stops in Penang, Singapore and New Caledonia.

The next day I set out on the first leg of my journey – an overnight run by train and ferry to the Island of Penang, Malaysia. Still very sick, I spent a few days in George Town and then began the airborne portion of the journey, the one that would put many miles between me and the source of my discomfort. The first leg was a quick morning flight from Penang to Singapore. As I had a half-day layover there, I decided to see the city, and found myself in a city museum.

A HANDFUL OF MOON ROCKS

What I recall of my time in Singapore is staring at length at a handful of genuine moon rocks that NASA had put on a world tour.

I was entranced, for these were the very rocks I had seen being collected, live on TV, by the Apollo astronauts who had landed on the moon over 220,000 miles away, less than a decade before. Now those same moon rocks were on display, inches from my hands. I didn't know it then, but some semblance of a theme had begun to take form for my life journey, one that would ultimately include outer space and better managing our planet's natural resources.

Returning to the pristine Singaporean airport later that day, I checked in and boarded a half-full aircraft. The extended range UTA DC-10-30 aircraft had begun its journey in Paris the day before. It was now bound for French New Caledonia for refuelling and would then continue southeastward toward New Zealand.

Soon after liftoff, in pleasant ignorance, we indulged in freely flowing champagne and caviar in the spacious aircraft. I had a window seat, with no one near me, and I watched the silent lightning flashes in the darkness far below, north of Australia's coastline. I experienced a profound comfort. I was beginning to feel better – much better. I could have stayed up there in that protective cocoon of darkness and pampering, with the huge turbofan engines purring away, for a very long time. I was truly enjoying a complete detachment from life and the seemingly endless troubles to be found on planet Earth.

I snapped out of my private reverie as we approached the Nouméa airport in darkness. The plane was being pitched and tossed with blasts of air and sheets of rain until we were finally landed and safe on the ground, taxiing our way to the terminal. We were asked to deplane. As the aircraft was being fuelled and serviced, I noticed one of the engine's cowlings had been removed, and I could see many flashlights probing its workings.

INTERLUDE IN FRENCH NEW CALEDONIA

We passengers were informed soon thereafter that the plane needed minor servicing. We were then promptly bused to a lovely, quaint French hotel in town, where private rooms awaited us. I was fine with that, now feeling a sense of adventure and pointed in the right direction.

The stay in Nouméa was expected to be brief, several hours at most. However, that afternoon the passengers were informed that the plane was still not ready and that there was an additional logistical problem – that of getting us back to the airport. An already broad river had risen dramatically with the storm. We would be overnighting in Nouméa.

Outside the storm worsened. The wind was gusting more strongly than ever and the rain was relentlessly lashing the outside walls of our hotel.

The next morning we were informed the aircraft was ready and we needed to prepare immediately for a return to the airport.

We piled into buses and soon encountered a not-insignificant "logistical problem": in order to get to the airport, we would have to cross the wide roiling river, which had become increasingly swollen with the onslaught of rain. The bridge was not even visible. We poured out of the buses and into several waiting and open flatbed army trucks. Miraculously, the truck drivers navigated us across the raging river over the submerged and invisible bridge, in driving rain. On the other side we finished our run to the airport and were relieved to be finally boarding our aircraft.

Thoroughly drenched, but comfortably seated, we heard the DC-10-30's turbines light up and begin their soft growl. The aircraft pushed back, and spontaneously as one, the passengers broke into a joyful cheer. Notwithstanding the fact we were all soaked with rain on our outsides, we were happy and relieved. Adding to the ebullient mood, there was an abundance of free beverages to be consumed, for the "medicinal" purpose of warming our insides. We taxied through the blasts of wind and rain to the runway's end, took a 180-degree turn, paused, and then the pilot gunned it. The growl turned to a roar and a dozen or so seconds later the nose began to lift. We soon were rising through and then above the chaos, headed toward the happy islands of New Zealand.

Landing in Auckland, and then stepping into the fresh temperate air, felt like coming home to a place I had never been before. I saw green grass and wanted to kick my heels and embrace the ground. Instead I trudged on with my waterlogged packsack through customs, then hitchhiked, hopped on a bus or three, then a ferry and finally, after a blur of uncounted days of travel from northern Thailand, found myself settled and nested in a backpackers' cabin in Nelson – a lovely town located on the northern tip of the South Island of New Zealand. Things were working out nicely and I was beginning to feel great.

SPACE COLONIES AND INTELLIGENT LIFE – MEETING THE VISIONARIES

It was in Nelson that a visit to a tiny bookstore yielded me two writings that, added to my serendipitous moon rock encounter in Singapore, were going to reverberate through a life journey that I could never have anticipated.

The first was a book titled *Intelligent Life in the Universe*, written in 1966 by I.S. Shklovskiĭ and Carl Sagan.[1] This detailed and uniquely collaborative work between a Russian and an American scientist explored in a very rigorous manner the likelihood of there being other places in the cosmos that could support carbon-based life as we know it. I was fascinated by the logic, analysis and conclusions of this work, and the passion of the scientist authors.

Reading it strengthened my lifelong fascination with Nature and outer space.

The second writing captured an interview with Jacques Cousteau by Space Colony advocate Stewart Brand.[2] The interview had been undertaken in 1976 at NASA Headquarters and was attended by astronaut Russell Sweickart, NASA's Deputy Administrator George Low and Jacques Cousteau's son Philippe. In the interview, Brand asked Cousteau why he was interested in the space program. Cousteau commented that the "health of the ocean has to be checked all of the time" and that he was seeking "an indicator of health." And, being the visionary he was, he was seeing the potential of space-borne sensors in supporting this objective.

Cousteau's comments resonated strongly for two reasons. The first was that as a practising biologist I had discovered first-hand how poorly we understood the ecosystems that we were supposedly "managing." A few years before, while working for the provincial government in the Skeena Region of northern British Columbia, I was flabbergasted and disheartened to discover the paucity of diagnostic tools and information we had on which to base our management of fisheries resources – in this case six species of anadromous (ocean-going) salmon and steelhead trout that numbered in the many tens of millions of fish moving in and out of the huge Skeena system and the Pacific Ocean. I felt that our management efforts, particularly with respect to the compilation of inventories of both fisheries and wildlife populating an area of just under two million hectares, were entirely inadequate. Notwithstanding the fact that we resource managers were required to be making "sustainable" harvest decisions around everything from fish to grizzly bears, we had lacked fundamental, order-of-magnitude information on the populations we were responsible for managing.

The elegance, potential coverage, consistency and cost-effectiveness of using space-borne sensors to aid in resource management, whether on land or in oceans or estuarine ecosystems, had first resonated with me then – and has never left.

The second reason Cousteau's comments resonated was my interest in exploring the possibility of identifying a "proxy" indicator of planetary health or sustainability.

Underlying this interest was my concern that the continued well-being of humankind and our furred, feathered and finned cohabitants was jeopardized on a planet pushed beyond its abilities to accommodate impacts and recover homeostasis. And how could we know the status of our planet without a measure or a suite of measures by which to assess it?

DR. KEELING AND MR. GORE – AN UNSETTLING DISCOVERY AT HAWAII'S MAUNA LOA OBSERVATORY

At the same time, on the Big Island of Hawaii, located between New Zealand and my home in Canada, a story was unfolding at the hands of another visionary scientist, whose fame would come later. Dr. Charles David Keeling, working for the Scripps Institution of Oceanography at University of California (San Diego), was quietly gathering information at the Mauna Loa Observatory on the Big Island of Hawaii that would have a dramatic impact on the way humankind looked at the world. This would result in trillions of dollars being redirected from conventional investments into a range of activities relating to renewable energy, carbon-financed forest management, sustainable agriculture, innovation and "clean" technology.

Keeling's work had begun quietly in 1958 and was focused on a regime of comprehensive measures of CO_2 concentrations in the atmosphere. As I travelled through New Zealand, Keeling was into his 20th year of taking regular measurements in the concentration of CO_2 in the atmosphere, notwithstanding the fact that the National Science Foundation stopped financing his CO_2 research in the early 1960s. Ironically, in 1963 the National Science Foundation used Keeling's work to caution about the potential of a CO_2-driven greenhouse effect, a position that was echoed in a 1965 report from President Johnson's Science Advisory Committee, which had warned of the possible dangers of such an effect.

Keeling persevered, and his measurements conclusively demonstrated that global atmospheric concentrations were on a steady rise. At the beginning of his work in 1958, the CO_2 concentration was approximately 315 parts per million (ppm). As I hitchhiked through New Zealand in 1978, the concentration had risen to 336 ppm; when I would be doing my research a decade later, it would be 349 ppm. This past year (2016) has been characterized by measures of atmospheric CO_2 that have consistently exceeded the 400 ppm level, and increased attention being paid to apparently abnormal and severe weather events from around the world.

These increases in CO_2 concentrations in the atmosphere were presumed to be primarily a result of growing anthropogenic (man-made) emissions arising from the combustion of various forms of fossil energy: primarily coal, oil and natural gas.

Keeling's second discovery was that the atmospheric CO_2 concentration curve had a significant annual ~5 per cent reduction beginning in May of each year, representing approximately 40 billion tonnes of CO_2 removed from the atmosphere. Keeling and his colleagues ultimately concluded that decreasing atmospheric CO_2 concentrations beginning in May of each year were the direct consequence of enormous CO_2 removals accomplished by northern temperate forests as they exited winter dormancy, "greened" and became photosynthetically active. Conversely, when the trees of the temperate forests stopped growing and slipped back into dormancy for the fall and winter, they stopped sequestering (removing and storing) CO_2 and switched into becoming net emitters of CO_2 through their respiration.

These natural emissions, combined with anthropogenic emissions, more than counterbalanced the removals, by about 4 billion tonnes every year.

With President Johnson's Science Advisory Committee report warning of the possible dangers of a "greenhouse effect" attributable to rising CO_2 concentrations, Keeling's work began to be watched carefully, and replicated around the globe.

As it turned out, Cousteau's illumination of the potential of space-borne capacity to monitor ocean ecosystems and his quest for a single indicator of ocean health, Keeling's revelations around atmospheric CO_2 on Mauna Loa, and the rigorous study of life in space from Sagan and Shklovskiĭ, would continue to echo and direct my personal journey for decades to come.

HOMEWARD BOUND

Back in New Zealand, I had hooked up with my high school buddy Kenny, who happened to be on a photographic mission on the South Island, and we spent the next few weeks camping and exploring. Somewhere on the west coast, Kenny continued south, and I headed east on my own. This culminated with a week or so at the YWCA hostel in Christchurch. I jogged the Avon River twice daily and was now fully recharged, both in body and mind. But I felt a strong tug home to complete undefined, unfinished business, having something to do with energy and the environment. I booked my flights, flew to Auckland and hopped on board the same UTA DC-10-30 I had boarded in Singapore over a month before. I was homeward bound toward Vancouver, with stops in Tahiti and Los Angeles.

On the Tahiti to Los Angeles leg I perused a newspaper and learned that the *LA Times* Track Meet was happening the evening of my arrival. I decided to take a pause and make a surprise visit to my friend Debbie Brill.[3] Debbie was an Olympic athlete I'd met at the University of Victoria four years earlier. She was a member of Canada's National Olympic team and would surely be competing in the high jump. The *LA Times* Indoor Track and Field Meet, held at the LA Forum, was sold out, but I somehow finessed my way on to the ground level via a media entrance. I didn't want to rattle Debbie with a surprise before she competed, so I kept out of sight and waited for her event to complete. Debbie won her competition, as usual, and retired to the lower bleachers to warm down. After a few minutes, I quietly stepped in front of her, waited for her to look up, and waved. It was a great surprise.

When the meet was finished, we headed toward the LA Forum's ground floor exit with the other athletes. But before we were able to exit, Debbie was approached by a very tall black gentleman, who chatted and shared pleasantries with her. This turned out to be the legendary basketball player Wilt Chamberlain. We then walked into the balmy LA night, cabbed back to the hotel to get cleaned up, go for dinner and share an evening of catching up on each other's adventures.

Another male Olympic athlete, who appeared to be importing Italian sports cars as a sideline, joined us after dinner. We raced around the freeways in and around LA at some ridiculous speed in his very hot Lamborghini, dropped in on a few bars, returned to the hotel and then said goodbye.

Debbie would show up again later, as the themes arising from the moon rock encounter in Singapore, my scientific meanderings in Nelson and a few unexpected but critical events continued to draw me down a serpentine path to destinations unknown.

Gone Fishin'

1979–1982

On returning to Vancouver and Victoria after my abbreviated trip to Asia, the illuminations and themes arising from my travels could have been forgotten, but they were not. They were, however, going to be parked for the next several years, as I turned to unfinished "fish business" in my home province of British Columbia. I like to think of this as having given the new interests and directions "soak time." I am reminded of a dialogue described by Canadian astronaut Chris Hadfield, in his book *An Astronaut's Guide to Life on Earth.*[4] As Hadfield was hurtling around the planet aboard the International Space Station at 27,724 kilometres per hour, he was chatting on the phone with Canadian rock icon Neil Young, who was riding in the back seat of his hybrid 1959 Lincoln Continental somewhere in northern California. The two men could hardly have more different career paths, and yet they had much to talk about. Hadfield had made quite a splash with his musical performances from the International Space Station, playing David Bowie's "Space Oddity" on a Larrivée P-01 parlour guitar handcrafted at the Cordova Street shop in Vancouver and purchased by NASA. The all-wood guitar was built with BC Sitka spruce, African ebony, and Honduran mahogany. As wood is produced by the cambium of trees with carbon removed by the atmosphere through photosynthesis, the guitar represents a tangible example of sequestered carbon. The Larrivée P-01 guitar was the third in space.[5] It was taken aloft on August 10, 2001, aboard the *Discovery* during the STS-105 space shuttle mission.

Chris didn't just cover other musicians, he wrote original music and performed it – on one occasion he had hundreds of thousands of Canadian students joining in. Somewhere in their discussion around songwriting, Young had cautioned against trying to write down a song before it had adequate time to finish and was mature enough to flow of "its own accord."

I wasn't writing songs, but these were big concepts, representing a dramatic departure from my thoughts about a life direction, and I wasn't anywhere near ready to run with them.

FISH WORK

And so I returned to my work with fish. As a boy I always had kept aquariums, but that was an aesthetic interest, not a scientific one. I was enchanted by the colours and beauty of the tropical fish I kept, and I was truly fascinated by their behaviour.

I came to be a fish biologist neither by design nor intent. It happened in the spring of 1972. I was in my third year of a Zoology major, and looking for a summer job. I saw a posting in a biosciences building hallway outside the office of Professor John Krebs, a UBC behavioural scientist who was looking for a student to do research on the great blue heron. It turned out that John Krebs was a close relative of Professor Hans Adolf Krebs, who was famous in science for the "Krebs Cycle" he is credited with having discovered in 1937 and for which he won the Nobel Prize in 1953.

The Krebs Cycle is a ubiquitous biochemical pathway that produces bio-available energy[6] for most of the world's organisms. It is also responsible for virtually all of the CO_2 produced by respiring organisms on the planet, far exceeding, for example, the annual CO_2 emissions from burning fossil fuels.

This phenomenon of biological CO_2 generation (and its removal) would become highly relevant later in my life. But for me as an undergraduate in university, the Krebs Cycle was just another seemingly useless piece of information to be memorized for regurgitation during an exam and presumably forgotten thereafter. I didn't know it, but this was in fact my very first segue into the world of carbon. I would never have believed it would become the dominating focus in my life for over two decades.

In any case, the notion of studying the great blue heron had grabbed my attention. I applied for the job and met John at his office. We agreed to meet at a nearby rookery on Point Grey so that I could see what the job entailed and we could determine whether it was a good fit.

I met John at the rookery site beside a stand of Douglas fir trees. Dozens of large heron nests could be seen far above, presumably built there as an anti-predation strategy. To do the job of observing the herons' behaviour, I needed to attach tree-climbing spurs to my boots, put a belt around the tree and back to my hips and walk vertically (as linemen do on telephone poles) to the viewing platform 25 metres above me. I donned the gear, jammed the right leg spur into the bark, pushed up and did the same with the left leg. Then the right again. At this point I was perhaps 2 metres off the ground.

My bird career was finished. My knees began to shake uncontrollably and I was paralyzed with fear. John laughed and said in his very proper Oxfordshire dialect, "Yes, that's what happened to me too." I took a few deep breaths and descended to terra firma.

Soon thereafter I saw another job posting on campus, this one for doing sonic tagging and tracking of cutthroat trout in a small mountain lake located in the UBC Research Forest near Maple Ridge. This required no more than the ability to row a small boat, point a submerged microphone to locate the fish through triangulation and then track their movements all night as the fish moved around. Or didn't move – which is mostly what happened. No shaking knees. In fact the major challenge was staying awake. This was a job I could handle. I was definitely qualified and I took it.

Thus started the fish biology part of my career. My fish work at UBC was followed with stints with the Marine Biology Division of the Royal British Columbia Museum and the provincial government's ELUC[7] Secretariat in Victoria, the Fish and Wildlife Branch in the Skeena Region and more.

ENERGY AND THE ENVIRONMENT

On returning to British Columbia after my trip abroad, my primary interest was in identifying the impacts of large energy projects on fish populations and fish habitat, primarily that of salmonids,[8] and the identification of impact mitigation options. These options included fish habitat enhancement and ecosystem restoration. At that time there was virtually no consideration being given to the potential impacts of energy projects on CO_2 emissions and climate change.

My first clients and employers included First Nations, the International Pacific Salmon Fisheries Commission, Environment Canada, BC Hydro, an advocacy group reporting to a Royal Commission in BC, and the Government of British Columbia.

I much enjoyed fisheries work, particularly exploring and assessing small coastal streams supporting coho salmon and cutthroat trout. The first task was to walk each stream from the ocean to the blockage to fish passage, which was usually a falls. Then we had to determine where the productivity bottlenecks were to be found, as this is where the enhancement and restoration opportunities were located. In some cases fish ladders could be employed to increase the rearing area, and in other circumstances the careful placement of spawning gravel was prescribed.

I was blessed with a great teacher and mentor – the irascible and enigmatic biologist, Ted Burns. Ted was from the East Kootenays, and apparently had some Native blood in him. He was a shy and gentle soul, and while soft-spoken, he didn't mince words. His eyes were clear and bright – and of two different colours.

Ted related to fish almost on a personal level, as if they were cousins, for example referring to juvenile coho salmon as "boneheads," as he considered them being light in grey matter. On the other hand I was an ivory tower academic who studied the esoteric, such as schooling behaviour of zebra danios and neon tetras in variable light conditions, in the comfort of well-equipped labs. I knew virtually nothing practical about managing native fish populations. But Ted showed me the ropes.

"Burnsy" also had a tendency to leak confidential reports to newspapers. These reports were usually prepared for government, and if he was concerned that his recommendations were falling on deaf ears or "disappeared," out would come the plain brown paper envelope. Ted took great pride in his work, to the point of hand drawing his enhancement structures, such as fish ladders, customized to each creek. He was intolerant of his work being ignored.

As these leaked reports had my name on them as a co-author, this practice no doubt affected my future employment potential, at least with the provincial government. But in terms of getting the truth out, I was with Ted and accepted the consequences.

I had a penchant for sports cars, and Ted's truck (sometimes known as a "Shaggin' Wagon") always seemed to be in the shop. So we would take my little Alfa Romeo Berlina sedan to the logging roads to do our fish work. We would treat the high output DOHC (double overhead cam) "Italian stallion" as if it were a Ford 4 × 4 pickup on steroids. We called it the "ATV." It was ridiculous, but we had a wonderful time accessing the badly impacted creeks of southeast Vancouver Island with the ATV and then going back to Victoria in the evenings to prepare plans for restoring each one of them. As noted, Ted's illustrations were all by hand – pen and ink. I did much of the writing, using a pink IBM Selectric. This typewriter represented the absolute state of the art in typing technology at the time.

Both of the fish species we were focused on, coho salmon and cutthroat trout, spawn in the small coastal streams, then go to the ocean where the food supply is much larger. The coho spend a year rearing in the stream, then go through the smolting process which turns them bright silver. Coho spend 6 to 18 months in the ocean, where they grow quickly from grams to several kilograms, and then return to their home stream to spawn and die, completing the cycle of life and death that all true salmon share.

Cutthroat trout are much more varied in their lifestyles. Some stay in the stream for their entire life, others "smolt" and go to the ocean and estuary for the richer food supply there, and then return to the stream to spawn[9] and reside. Or they go back to sea. Cutthroats were nomadic. No steadfast rules. No lifestyle bible to abide by. I liked them.

When doing inventory work on the large river systems and lakes of northern BC, if they were inaccessible by road, we would fly to the site by fixed-wing aircraft (primarily de Havilland Otters and Beavers) or a Bell Jet Ranger helicopter. In doing so, my workmates and I experienced BC's raw and spectacular scenery and wildlife from a unique vantage point. It was a genuine privilege for which I am thankful.

In some circumstances, the fish jobs would flow from one to another seamlessly. One such case had me doing a solo "all-nighter" monitoring water chemistry and other parameters on

Nicomen Slough near Mission. This work was on behalf of a First Nations group considering the potential of the study area for establishing an aquaculture operation for carp. When morning arrived and my final dataset was complete, I loaded the red Volkswagen van I then owned and drove home – only to immediately set off for the next mission. This job involved working with the International Pacific Salmon Fisheries Commission (IPSFC), an international body having shared responsibility with the Canadian Department of Fisheries and Oceans for the management of the Fraser River's salmon.

As the Fraser River's salmon stocks "far exceed any other river on this continent, or for that matter, any other river in the world"[10] it was going to be an honour and challenge to work on this extraordinary waterway. I was very thankful for the opportunity.

Without sleep, I picked up a Fisheries Commission vehicle in New Westminster, loaded my gear and headed north to Lillooet by mid-afternoon. I had been given directions and a key by the lead IPSFC biologist so that I could overnight at a Fisheries cabin that would be my accommodation for two or three days. It was located about six kilometres west of Lillooet. After several hours of driving through the Fraser Valley and Canyon, I arrived at the cabin. It was well after dark and there was no power or light, but I managed to open the padlock and make my way into the tiny cabin. I stumbled around and lit a candle. Then I threw my sleeping bag onto a cot. I was completely exhausted and immediately fell into a very deep sleep.

In the middle of the night, my sleep was shattered by sounds – and this time it wasn't chanting monks I was hearing. There was no light, and I didn't know where I was. I was completely surrounded by a loud and eerie cacophony of yipping and yapping. It sounded and felt as though a pack of wild animals, which had to be coyotes, was inside the cabin. Of course the animals were not inside, but without question they were only metres away from me. The howling carried on for ten or so minutes and then stopped as abruptly as it started. They were gone.

The next morning I met the IPSFC biologists who explained their work in Lillooet, which was largely directed to solving a "riddle" involving sockeye salmon migrating to their nearby spawning grounds. Like coho, sockeye salmon return from one or more years in the ocean to spawn in the stream or lake in which they were born. This journey, which takes many months, begins in the open Pacific Ocean and can entail thousands of kilometres of travel inland.

The sockeye we were focused on were trying to return to spawn in their native Cayoosh Creek but were being attracted to the turbulent outflow from a hydro generation facility on the Fraser River, instead of the creek which entered the Fraser upstream. Why the confusion? Simple. Sockeye are equipped with exceptional olfactory equipment, and on their spawning migration find the place of their origin by using their advanced sense of smell. The hydro facility's water was being sourced from Cayoosh Creek, which entered the Fraser upstream of the hydro generation facility. However, on their upstream journey, the sockeye encountered the discharge from the hydro facility first, as it was positioned downstream of Cayoosh Creek. The scent of their home stream was coming from the facility, so that is where they stationed, and then tried to head upstream.

Predictably, the sockeyes' efforts to swim through the outfall and into the plant's turbines were not going well. Hundreds of sockeye were forming a giant and growing gyre of fish, turning slowly like a dark underwater galaxy, at the foot of the hydro facility on the Fraser River.

Consequently, BC Hydro and the IPSFC were co-operating in trying to correct this problem and were radiotagging the sockeye to determine how their efforts were faring with the migrating fish.

I worked alongside the IPSFC biologists for a few days, learning how the fish radiotagging and tracking equipment worked, which I would need to know for my real job, which was awaiting me in Quesnel, BC.

RHYTHM OF THE RIVER

Finished in Lillooet, I had my last breakfast at the Sasquatch Inn, and still wearing an IPSFC "hat," I drove north for the better part of a day to Quesnel. There I joined a team of consultants who were working on a study informing the Kemano II Expansion project, a proposed hydro facility near Kitimat. This was a major hydro project, involving the redirection of Fraser watershed water arising from the northern interior of BC, through a giant tunnel drilled through a mountainside to the coast – and a large power generation plant. Most of the energy would be used to smelt aluminum in Kitimat. This water would under normal circumstances join the

Fraser River and enter the Pacific Ocean near Vancouver, located 655 kilometres south of Kitimat.

This was no minor project, nor would be the impacts. Biologists were concerned that the diversion of so much water from the Nechako River, a Fraser tributary, would have a significant impact on river flows and temperatures, and thereby affect spawning salmon and other species. As part of a study to secure baseline information and investigate potential impacts, the team was planning to radiotag sockeye salmon and then follow their migration up the Nechako River. For this project they had secured a Cessna 170 fixed-wing aircraft, a Bell Jet Ranger helicopter, seine nets and much electronic paraphernalia that they hoped would allow them to capture, tag, release and track hundreds of sockeye up the Nechako and Fraser Rivers.

There was a problem. Notwithstanding the big budget, the choppers, seine nets and more, the tagging team could barely catch a fish. The plan, presumably hatched in some downtown Vancouver office, simply did not work. Exasperated, they eventually resorted to borrowing "ramps" from some of the local fishermen[11] who knew how to catch fish most effectively, although not necessarily legally. Each ramp consisted of a wooden frame walkway that extended from the shore at right angles, out three metres or so into the river. Chicken-wire mesh was attached to the frame, top to bottom, so the sockeye migrating up the river's edge would bump into the mesh, then swim obliquely to deeper water and the end of the ramp. Here they would turn the corner and swim upstream. Or at least try to.

Perched at the end of the ramp would be the fisher, waiting for the unseen fishes' attempts. The fisher would be facing upstream, and stroke a "dip net" from upstream to downstream with a paddling motion, as if paddling a canoe while standing, seeing nothing, hoping...

This fishing system only worked at night. We would take 10- or 15-minute shifts, with a half dozen of us rotating responsibilities through the night.

Every once in a while there would be a shout and some splashing. The onshore team members would snap to attention and secure the netted fish, carefully insert a radiotag through its mouth into the fish's stomach. At the same time, measurements and gender were quickly recorded. Then the fish was released back into the river, where it would continue its sojourn toward its home stream. The crew was practised, and the entire operation might have taken 20 seconds before the fish was swimming upstream again.[12]

During the day, fixed-wing aircraft fitted with radios and antennae followed the fishes' progress as they migrated farther north toward the spawning grounds. It was hoped that the timing, migration speed and other information gathered could ultimately inform plans to mitigate impacts of the proposed project.

This was midsummer, and we only worked on the river at night, which made for some magical evenings. One lovely experience I had was hearing, then feeling, changes in the rhythm of the river's flow, now and again, throughout the night. The Fraser is a very large river, extending from the Rocky Mountains to the very large estuary where it enters the Pacific Ocean near Vancouver. The river was perhaps 60 metres across at our site and was clearly audible. What I will never forget is the sound of an approaching surge. One could hear it developing upstream, as the river picked up tempo, and then feel the surge as it hit the ramp. This would cause the entire wood and chicken-wire structure and its lone occupant to shudder for several moments.

Then it would settle back into its normal state. Occasionally, during a particularly strong surge, a massive boulder could be heard rumbling along the floor of the river, seeking its next resting spot.

In darkness, our senses were heightened. Once in a while the fisher would capture a spring salmon (or "chinook"), which were also migrating but in much smaller numbers. Even though these were healthy living fish, the distinctive smell of the spring salmon was unmistakable even from the shore – in fact I can still smell it.

OF MONSTERS AND MEN

Occasionally there was a monster in the middle of the night. On one occasion, the on-duty fisher, on making his paddle stroke with the net, experienced a resounding thump. This certainly wasn't any salmon. It felt more like a log – a log that was headed upstream. The fisher had firmly attached his net to the front end of a large white sturgeon,[13] lumbering upstream in the night, perhaps headed to the salmon spawning grounds to gorge on roe. White sturgeon can weigh over 1,500 pounds and grow to be 20 feet in length. They are long-lived, reaching sexual maturity at about the same age as humans. The white sturgeon, which first appeared in the fossil record 175 million years ago, is the largest freshwater fish in North America

Thankfully the fisher let go of the net's handle. In the moonlight, the crew stared in disbelief at the net handle, vertical and looking like a submarine's periscope, as it made its way upstream along with the ancient fish below.

Behind us was a riparian forest. Much life was packed into that zone, with many other sounds to be heard. Coyotes were heard regularly, with their incessant jabber. But the most startling sound we heard in the blackness was the scream of a rabbit, likely fallen prey to, and in the lethal talons of, one of the great horned owls that hunted there, on wings oh so silent.

At first light the sockeye either stopped moving upstream or saw the apparatus and avoided it. In either case the night's work would end. We would pack up and head to town for some sleep. Or we should have, but fish biologists and technicians of both genders are unpredictable and not easily contained by societal norms. My mentor Ted Burns was but one example.

After my first night of working the river, on reaching the motel room where several of us were staying, I was rather taken aback to see some members of the crew preparing martinis, "shaken not stirred." Actually, they were neither. They were either pure vodka, or more likely, lab ethanol, which like vodka is made from potatoes but has more than twice the alcohol content. After a night of camaraderie on the river, I felt obliged to join in for at least one, pleased to be working with this team of competent, dedicated, but off-the-wall biologists and technicians.

It was a rich and unique experience and I slept like the proverbial dog thereafter.

UNDER THE BOOM

Another fishy adventure during this interlude, not so pleasant, involved scuba diving in Tahsis Inlet, on the west coast of Vancouver Island. The area was very active in terms of forest operations. Fallen timber was taken by truck from the mountainsides to booming areas located on an ocean inlet. There the logs would be dumped into the ocean, sorted, "scaled,"[14] boomed – and hauled by tug out of the inlet to the open ocean and then to a sawmill or paper plant, of which there were several on the mainland coast and Vancouver Island.

To their credit, the forest operators in Tahsis wished to understand the impact of booming on the substrate ecology beneath the booms and had retained Ted Burns and me to take an initial look. Ted didn't trust technology much, and scuba not at all, so I "volunteered."

It wasn't pretty. Diving in the cold ocean water, under blackness and the impenetrable cover of a log boom, was unsettling. All I recall is the water getting darker and darker as I dove deeper and deeper. I was relieved to finally strike bottom. But it wasn't bottom. The inlet's floor was covered by a thick layer of bark and wood debris, burying whatever ecosystem had been established on the substrate. All I could see was the top of the debris layer. Surely this was not normal or natural to the area, or good for the substrate's native ecosystem.

I took notes and made my way to the surface at the edge of the log boom, then hauled myself out like a seal. I shared my observations with Ted, who recorded them and ultimately completed the work.

After packing the scuba gear into the truck, I grabbed myself a fishing rod and headed to a deep jade-green pool on the Tahsis River nearby. Almost immediately I hooked into a gorgeous, bright, summer run steelhead – one of two steelhead I had (and have) caught in my life. I played the steelhead carefully, unhooked it and then watched with pleasure, heart pumping and hands shaking, as the large silver rocket shot back into the dark green depths with one flip of its tailfin.

The adrenaline washed through me, and I was alone on the river. Or so it appeared, but given the area, I knew it was more likely than not that I was being observed, perhaps followed by one of the many cougars that preside over the Tahsis Inlet and the valleys behind it. Sometimes you can simply feel their presence – if not, the hair standing on the back of your neck may be the hint that you are being watched. Science hasn't adequately explained that one yet, but then, that is just one item in a long list.

FISHIN' CLOSURE

I continued my energy and fish-related work for a few more years, taking on a few other minor consulting gigs, including one relating to a proposed BC Hydro power-line alignment through the Lillooet River valley. However, the fire was no longer in my belly. I was being tugged in new directions, connected to those "parked" illuminations around carbon and outer space arising from my travels several years before, and I was more than ready to let go and reach for something fresh.

It was coming. And there was no way I could have guessed where *it* would take me.

Playing in God's Kitchen

1983–1990

I enjoyed my work with fisheries immensely. The unique characters I worked with, the extraordinary places I worked in, and of course the fish themselves and their remarkable life histories – all represented to me that which is priceless.

However, an echo from the not-distant past was tugging me in a different direction, and I was definitely ready for a new challenge. In 1983 I decided to return to my alma mater, the University of British Columbia (UBC), to begin graduate studies. My original intent was to undertake a master's program in Bio Resource Engineering. My interests were in aquaculture and the design of biological systems to deal with food production and waste recycling for space travel.[15] I also wanted to explore the potential for applying rapidly evolving space-borne remote sensing capacities to support natural resource management on Earth.

NORTH TO ALASKA AND A UNIQUE JOB POSTING

In the summer of 1983, my friend Jan, who was visiting from Switzerland, and I decided to take a road trip to the Yukon and Alaska. We rented a vehicle from a "discount" rental company, a gargantuan blue 1970s-vintage Chrysler New Yorker four-door sedan with a very large engine that sported eight cylinders but ran comfortably on seven.

Halfway to the Yukon border we discovered that three of the tires were essentially bald, and of different makes. Growing nervous, we examined the vehicle documents and found it was not insured. Jan negotiated with the rental company by payphone and secured some tires and insurance so that we could continue.

When we reached Whitehorse we surprised my great high school friend, Dan Davies, who was working as a fisheries biologist with the Yukon territorial government, for a two- or three-day visit. One morning, while I was enjoying some freshly brewed northern roast coffee and peering out the window at the Yukon River, Dan walked to the kitchen table with a folded newspaper in his hand. Dan had just happened to see an ad in a Whitehorse newspaper that morning seeking candidates for the "Canadian Astronaut Program." Knowing my penchant for outer space, he placed the paper, with the ad circled, in front of me over breakfast and smiled mischievously.

The posting, which represented Canada's tentative first step into the space program, stated that the Government of Canada was preparing to participate with the United States in future space shuttle missions and was calling upon candidates with suitable backgrounds to be trained to become "Payload Specialists."

Candidates needed to meet appropriate medical requirements. Practical "experience in flying," while not essential, would be an asset.

I had no flying experience,[16] but I saw the potential for my new academic focus on food and waste systems design to be relevant to the astronaut program's interests and decided to go for it. I fashioned my story, secured the support of a senior botany professor on campus as a reference, submitted a letter application on August 3, 1983, and in September I went back to studying in earnest for a master's degree in the Bio Resource Engineering Department at UBC.

GAME ON

To my surprise and delight, later that month I received a letter back from the Canadian Astronaut Program, over the signature of Mr. R.W. Dolan of the National Research Council of Canada. The letter stated that out of 2,400 applications, mine was one of those chosen for further consideration. There would be a subsequent process to confirm the final candidates chosen for training.

I was elated and energized, so I decided to get myself prepared for the challenge. I focused my interests at UBC on the design of biological systems that would support food production and biological waste management in space travel, with aquaculture as a backup.

My personal race to space was on, and it was exhilarating.

I quickly focused on fitness, which I knew would be a key criterion of any qualification process. I was fortunate to have as a friend Debbie Brill, the Canadian National Olympic team athlete I had met at the University of Victoria in 1974. I asked Debbie if she would help me get in shape, and she quickly wrote up a program. For the ensuing months, I cycled and trained virtually every day, with the hope, and the occasional apprehension, that my opportunity to do the unthinkable – as in blast-off from Earth on the controlled explosion known as the Saturn 5 rocket, and then orbit planet Earth – would become a reality.

In mid-October, a second letter arrived from the Canadian Astronaut Program. A number of finalists had been selected. I was not among them. I was disappointed but not surprised. While no reasons were given, there were inescapable realities involved. The facts that I had no flying experience, I was not bilingual, and I had not completed my graduate work were no doubt among them. I simply was not ready to enter the program. Dr. Marc Garneau, who already had combat experience in the Canadian Armed Forces and a long list of other qualifications, clearly had "the right stuff," received the training and carried on to fly in three space shuttle missions: STS-41-G, STS-77 and STS-97. Garneau is now a Member of Parliament from Quebec and the federal Minister of Transport.

POSING A QUANTUM PHYSICIST THE HARD QUESTIONS

My interest in outer space, piqued years before in New Zealand, was far from extinguished; in fact it was growing. In my spare time, I was playing with some fundamental questions around physics and cosmology. This had nothing to do with academe, at least not for me. Eventually I got "stuck" on a rather esoteric problem of physics and sent a letter to Professor Freeman Dyson at the Institute for Advanced Studies in Princeton requesting help in solving it. Freeman literally "wrote the book" on advanced quantum physics and was a world-class mathematician with a lifetime appointment at the institute. An accomplished author, having published several popular books,[17] Freeman was also a designer of nuclear-powered spacecraft[18] and had worked with an array of highly accomplished scientists.[19]

Freeman pointed out in his first letter[20] to me that I was asking two of the fundamental questions of the Universe. One pertained to "dragging of the inertial frame," and could be answered "if one believes in Einstein's equations."

Freeman went on to say that the second problem I had posed, which pertained to the behaviour of a spinning planet in an empty Universe, was "much more difficult" but was possibly addressed by Mach's Principle. Freeman noted that the principle had "never been formulated in a mathematically precise form ... and may never find a final answer."

A fundamental message arising from Mach's Principle is that every piece of matter in the Universe is connected to every other piece of matter in the Universe, and that this connectivity helps shape planets and define their movements. Apparently, formulating this connectivity mathematically has continued to prove elusive, but I seem to recall more than one poet having vividly expressed the connectedness.[21]

In retrospect, this was heady stuff for a fish biologist. But I was motivated and genuinely honoured to be in a dialogue with an experienced scientist and author with the intellect, integrity and wisdom of Freeman Dyson. Notwithstanding my place at the bottom of the cosmology totem pole, I had obviously posed some of the right questions to the right person.

It was a lonely time. Not one person, save Freeman, had any idea what I was working on or focused on, beyond the requirements of my graduate program.

ALEUTIAN CANOES AND DESIGNER TREES

As it turned out, Freeman and I had some common interests beyond cosmology. His son George[22] lived in a treehouse in Belcarra Park near Vancouver, building baidarkas[23] (large ocean-going kayaks originally of Aleutian design) for travelling the same BC coast that I grew up on, and that I had continued to explore by more conventional means. I visited George at his home in the park. I recall posing a question on some aspect of buoyancy I was trying to understand. He paused, then looked at me and said that it was a differential calculus equation. And so while his personal interest was Aleutian canoe designs, a large measure of mathematical skills seemed to be hardwired into George. Or perhaps that competency came with having had the "Father of the H-Bomb," the brilliant and iconic theoretical physicist Edward Teller, as a babysitter.

Amazingly for a world-class quantum physicist who designed spacecraft, Freeman had an interest in trees – including their potential role in carbon sequestration and possibly climate mitigation. Freeman was one of the first scientists to look carefully at climate change. The world was catching up to him: climate change and carbon management were increasingly being discussed, and the potential roles of trees and forests were going to be explored. Freeman had also been considering the re-engineering of trees and their photosynthetic mechanisms to improve their ability to deliver valued products and services to mankind.

In his handwritten reply to my letter, Freeman said, "My favourite scheme is to grow trees which photosynthesize liquid fuel instead of wood. The fuel then flowing down to a living pipeline underground. The trees then need never be harvested and the valley can stay permanently as pleasant as the Muir Woods."

We traded ideas around trees, and their redesign,[24] and then I went back to work on my graduate degree.

"YOU HAVE A PROBLEM..."

One day during the first semester of 1985, a senior UBC professor took me aside to engage me in a very important discussion. He shared with me his opinion that under my current supervisory circumstances,[25] I would never graduate with a master's degree. He asked me to take a pause and give very serious thought to precisely what I wanted to do in graduate school. I did as asked, consulted with a confidante and came to my decision.

We met again, and I told him I wanted to do my PhD. He asked if I was certain, and I told him I was. The focus would be photosynthesis, wood formation and the source-sink relationships linking the two phenomena.

We worked out a simple agreement. I would enroll in a master's program in Resource Management Science. One year into the program, I would challenge the comprehensive oral exams for the PhD. If I passed, I would immediately be transferred to a PhD program that was interdisciplinary between Botany in the Faculty of Science and Wood Science in the Faculty of Forestry.

And that is exactly what happened. My work was on the white pine tree native to northeast BC. I would explore and elucidate on the relationship between photosynthesis, which harnesses solar energy to remove CO_2 from the atmosphere, and the "cambium," a cylinder of embryonic tissue located under the bark, which takes the carbon bound in sugars and other chemical compounds and makes wood with it.[26]

It was a fascinating project of discovery that required me to learn how to operate an electron microscope to examine wood tissue that was sputter-coated with gold, and operate a sophisticated Kontron image analyzer that took tens of thousands of complex cell dimension measures[27] per minute. I also learned to manage an array of other diagnostic equipment to measure ambient CO_2 concentrations, CO_2 uptake and other parameters.

"IT'S ALL AROUND IF WE COULD BUT PERCEIVE":[28] SEEING THE TREES' GLOW

Of these instruments, the most intriguing to me was the "Integrated Plant Fluorometer." This instrument could "see" the reddish glow (termed "fluorescence") that is given off by the antennae pigments in the chlorophyll found in *all* photosynthesizing plants – from gigantic redwood trees to microscopic ocean plankton.

This glow, too dim to be visible to the human eye, and only happening in light, has a great deal of information packed into it. Nurseries used the fluorometer to identify and weed out sick trees, which emitted differently than healthy ones, before they were planted – and well before any malaise was visible to the naked eye.

The application I developed was to use a variant of this fluorescent glow, termed "O-level fluorescence," as a "signature" to determine on a real-time basis how fast a particular tree was removing carbon from the atmosphere and making it into wood. I was at UBC, but the fluorometer and CO_2 measurement instrumentation were at Simon Fraser University (SFU), so each time I was to measure my trees I would pack them into a van, like a well-behaved peewee hockey team, and drive them to the SFU campus in Burnaby.

The seedlings were dark-adapted for five minutes and then tested for fluorescence under a blast of light at a controlled intensity and frequency. The fluorescence for each tree was measured 1,000 times per second for five minutes, and the data were logged for future analysis. As part of a broader protocol, we measured atmospheric CO_2 levels, which in 1989 were approximately 349 ppm.[29] I had no idea these measures would become so important a factor in my future, and ultimately that of the planet.

After the battery of measurements, I would truck the trees back to UBC.

Analyzing the data later the following year, I was very excited to find a very strong correlation between fluorescence measures and the rate of wood formation, determined using the Kontron image analyzer, an electron microscope and mass balance measures. These findings would be confirmed with my successful thesis defence in February 1990.[30]

SQUARE TREES AND PLAYING IN GOD'S KITCHEN

While I was still working on my thesis, driven purely by intellectual curiosity, I set out to understand how it was that the stems of some tree species manage to grow eccentrically[31] in cross-section in a sustained wind environment. Armed with a growing understanding as to how wood formation was controlled by various biophysical factors affecting the cambial tissue that forms wood, I played with some ideas, landed on a theory and decided to try it out. The challenge was to artificially promote eccentric tree development, without wind, in the laboratory.[32]

To my surprise, not only could I make trees grow eccentrically in the laboratory but I was able to coerce them to take on other shapes as well. After hearing about my work, (the late) Laszlo Paszner, a wood science professor with the UBC Faculty of Forestry whose interests were largely shaped by commercial considerations, challenged me to grow trees in a "square" cross-section. The context for this challenge was the fact that large amounts of waste are generated in the process of converting round stems into rectangular dimension lumber at sawmills. Also, the loading of raw logs onto trucks would be much more efficient if the logs were "square."

I really did not care about commerce at that time, but taking on the intellectual challenge was another matter. Could I outsmart or trick the trees? I took on the challenge, notwithstanding light-hearted cautions from Professor Vladimir Krajina. Vladimir was a celebrated Order of Canada ecologist[33] who had been a leader of the Czech resistance in World War II and had escaped the Gestapo many times.

With eyes that sparkled like those of Jimmy Durante in the limelight, Vladimir had cautioned me of the dangers of "playing in God's kitchen." But I noticed he was smiling.

First an associate developed a computer model that demonstrated how over time an initially round stem could transition to have a visibly square-ish cross-section. In the computer model, this was achieved by accelerating the rate of cambial activity (wood formation) where the corners would be. But could we duplicate that with a real tree and make it grow square-ish?

I applied some newly discovered insights into how wood formation was controlled, and sure enough, I could trick the trees into taking on square-ish cross-sections, consistent with the computer growth model. News of this discovery was first reported when I was interviewed on a national CBC Radio program in 1988. This was followed by a number of national network television interviews. Thereafter the phone in our lab started ringing frequently, with science and business reporters calling from around the world, wanting to get the scoop on the quirky development. One gentleman, who had forest land in Costa Rica, drove out from Calgary to see the square-ish sequoia tree I had growing on my porch at home, wanting to know if we could add value to his teak plantation by making the trees grow square.

Stories began appearing in a range of publications, including *The Globe and Mail*'s Report on Business,[34] and the venerable London *Economist*, which took an interest and published a square tree story on April 15, 1989 while I was immersed in thesis preparation.[35]

And, unbeknownst to me, the square tree story had caught the interest of a well-published author whose writing I admired. I learned of this years later when, one evening over a beer, a friend asked me how my square-ish trees had made their way into *Jurassic Park*, the 1990 novel written by (the late) Michael Crichton which provided the basis of the popular film of the same name. This came as a complete surprise, which I questioned, but my friend presented the paperback, opened to page 1, and there it was, the reference to "square trees tailor-made for lumber."[36]

TREES AND CLIMATE CHANGE: JOINING THE DOTS

On the more serious side, in the midst of the continued flurry of media attention on our lab, I took one phone call that was going to "join the dots" for me and which foreshadowed a career path I had never contemplated. The call came from a science reporter at *The New York Times* who had heard the news and then tracked me down, seeking the "secret formula" for making trees grow square. I avoided disclosing specifics of my technique, as I had one related patent pending and another under development. And more importantly, I was having a great deal of fun with the "secret."

The reporter was trained in science and had quickly realized that more wood formation translated into more CO_2 removed from the atmosphere, and that if this accelerated growth (not necessarily square) and photosynthetic activity could be scaled up in large forests, it could have implications for mitigating climate change. This assumed global mean temperatures had a causal relationship with atmospheric CO_2 concentrations. The reporter shared this observation with me. I had to concur, then mentally filed away the discussion and got back to working in the lab. If a story was published, I did not see it.

From that point forward, I saw the tree from a completely different perspective, more as a "solar-powered carbon removal unit" that had the added virtues of making the removed carbon into a solid fuel (wood). Considering the dilute atmospheric concentration of CO_2 of about 350 ppm (at that time), trees were very efficient at doing this. For example, a one-acre cluster of such "solar-powered carbon removal units," otherwise known as a "forest," situated in coastal BC, could in one year remove the equivalent content of CO_2 found in a one-kilometre column of air above it.[37] No existing or anticipated technology could even begin to match this capacity to remove CO_2 from the atmosphere.

And the forest would do so at negligible operating cost, with the added benefits of providing wildlife habitat, storm buffering, edible plants, fuel (wood), animal protein and mushrooms for those practised in culinary arts. Forests and trees began to take on a new status for me.

Surely, if a human had invented such a miraculous gizmo, she or he would be considered a genius and awarded the Nobel Peace Prize.

However, like the incorrigible Rodney Dangerfield, trees just "don't get no respect."

A SURREAL END TO ACADEME

My academic experience ended oddly – with a somewhat unique piece of drama that defies any explanation. But I will share it.

Having missed the convocation ceremonies when I graduated with my undergraduate degree in zoology 16 years earlier, I decided to don the full regalia, including the flat black hat and cape trimmed in burgundy, and say goodbye to the academic life in style. The end was nigh. But I never could have anticipated that it would play out as it did.

The formal ceremony was held in UBC's War Memorial Gym, and I was one of thousands called to the stage, to receive their due recognition. I was a tad nervous, and almost walked past the chancellor – *sans* my degree. He hailed me down, smiled, said some kind words, bonked me on the head, or shoulder, and off I went. My life as a graduate student was over.

Given the circumstances, I thought it apropos to go to the lounge in the Student Union Building and celebrate this final goodbye to academe with a cool beverage. It was early, perhaps 4:00 p.m., and only one person was sitting in the lounge – a svelte and well-dressed young brunette, perhaps in her early thirties. I sat a few tables away and ordered a lager from the waitress. Somehow the young lady and I started chatting. I was still in my full regalia, so we discussed the convocation and then carried on with more small talk.

At some point she asked for her bill and presented her charge card to the waitress, who disappeared behind the bar area. The waitress seemed to take an excess of time, as it was not busy, but my new acquaintance was not in a hurry. We continued chatting, as I noticed two burly RCMP officers approach the lounge. They entered, talked with the waitress, then strolled over to the young lady I was talking with. I couldn't hear what was said, but after the police approached her, words were exchanged. She stood up, and without looking back or saying anything to me, left the Lounge with an RCMP police officer on each side. The three disappeared down a hallway, reappeared a few minutes later, and then the officers escorted her out of the building, one on either side of her slender frame.

I asked the waitress what was up, and she replied that she had a "funny feeling" about my acquaintance, called in the credit card and found it was fraudulent. And so she called the campus RCMP.

It was a surreal ending, but thus was my graduation celebration on leaving UBC, having received my PhD at the convocation, after a total of 12 years on campus.

I removed the hat and gown, returned them to the bookstore where I had rented them and drove home to my place in the West End of Vancouver. I needed some soak time. The happy twittering of my pet zebra finches was the perfect elixir for easing out of one incredible era, of playing in God's kitchen, to prepare for the next, which would have me landing in the proverbial flame. Of what was coming, I had absolutely no idea and wouldn't have believed it if I'd been told.

Into the Fire

1990–1997

AN END TO UNFETTERED INDUSTRIAL OPERATIONS

Within a year of the *Economist* article, I had defended my PhD. It was time to leave the comfort and safety of academe's Ivory Tower – and so for the next chapter in life I decided to dive into the fire of industry. Things were heating up.

By 1990, any industrial executive or officer with an ear to the ground knew that the era of relatively unfettered industrial operations was coming to a close. Non-governmental organizations (NGOs) were becoming increasingly active, often with the media in tow. In Alberta, the Pembina Institute, which had its beginnings in 1985 as a Canadian non-profit think tank focused on energy policy, was going to become a significant player and watchdog, particularly with respect to climate change. In Ottawa, the Sierra Club, with its new leader Elizabeth May, was ramping up its presence.

In British Columbia, David Suzuki, who was somewhat of a celebrity for his role hosting the CBC programs *The Nature of Things* and *Quirks and Quarks*, was quietly choosing a board of directors for the Suzuki Foundation in his Vancouver living room.[38] David then launched the organization in early fall of 1990.

In short, companies were facing both NGO attention and pollution control regulations that had significant costs associated with them. There were a few high-profile court cases that had the attention of the executive "suits." Perhaps the most significant was the Province of Ontario's court case against Bata Industries Ltd.

Following a surprise visit by Ontario regulatory officials in August 1989, Bata Industries Ltd., a well-established shoe manufacturer, was charged under the Ontario Water Resources Act and the Environmental Protection Act with operating a non-compliant waste management system, causing adverse effects on the natural environment. Four of Bata's corporate officers and directors were charged with failing to take reasonable care to prevent this discharge and with failure to notify the appropriate authorities. This was one of the first high-profile cases in Canada where corporate officers and directors were individually charged and faced fines and imprisonment for environmental offences committed by their company.

The Bata incident demonstrated the potential consequences of not having an appropriate "Environmental Management System" in place to address environmental responsibilities in a professional and scientifically defensible manner, meeting regulatory requirements. As the new enforcement provisions could send executives and directors behind bars, government regulators finally had the full attention of the "suits." A new era of regulatory rigour was approaching at lightning speed.

For most environmental issues, such as the release of waste streams into water bodies and air sheds, technological solutions were available. The deployment of "Best Available Control Technologies" (BACT) or some variant thereof, was increasingly becoming a regulatory requirement before permits would be issued for constructing and operating new industrial facilities, such as power plants, refineries and manufacturing plants.

While capital intensive, BACT applications represented "the easy stuff," as the regulations were prescriptive. In the case of regulated energy utilities, there was no problem, as the costs of regulatory compliance were built into the rate base and passed on to the customer. The regulated company could actually earn a return on its pollution control investment. But a much more challenging issue was gathering steam.

A NEW ISSUE EMERGES

Since 1987 the Montreal Protocol on Substances that Deplete the Ozone Layer had begun to put a focus on "radiative forcing" in the atmosphere, otherwise known as the "greenhouse effect," and the writing was on the wall concerning an emerging environmental issue for which there was no BACT solution. The issue was climate change, and the primary substance that was going to be targeted was carbon dioxide (CO_2).[39] While scrubbing CO_2 from an industrial waste stream is easily accomplished using catalysts, there existed no practical, cost-effective technology for disposing of the CO_2 removed from waste streams, or stripped from the raw natural gas, which in some areas was over 30 per cent CO_2. And there was no regulation requiring CO_2 to be controlled. Consequently, the gas was being released from facilities directly and invisibly to the atmosphere. With no practical means of disposing of CO_2 removed from industrial waste streams, no regulations controlling CO_2 releases and a steadily growing production in most of the world, emissions of CO_2 to the atmosphere were on the upswing.

INTO THE FIRE

It was within this new world of dramatically increased attention to environmental issues that I accepted a position as Environmental Coordinator with Canada's largest integrated natural gas company, Westcoast Energy Inc.[40] The job required oversight of impact assessments for over a billion dollars in planned new pipeline and gas plant facilities – at least that is how it started. I was excited but apprehensive, as I had always regarded "big energy" as being a bad actor, but the people I met in the interview process were professional and competent and I felt a strong pull. I also had a sense that there might be an opportunity to get creative with respect to addressing regulatory compliance, and perhaps even taking things a step further, achieving environmental performance that exceeded compliance requirements. I was pleasantly surprised that the executive offices were open to taking such a route, within reason of course.

Learning about the natural gas industry by studying reports at my desk, hidden in the corner of a small reference library, I came to understand that the raw natural gas the company was gathering and processing in northern BC had mixed within it significant quantities of H_2S and SO_2, as well as propanes, butanes and CO_2. All of these impurities had to be removed at the gas plants, so that the company could ship "sales gas," virtually pure methane (CH_4), through the company's pipelines to markets as far south as California.

The sulphur contained in H_2S, a lethal gas, and SO_2, many thousands of tonnes per day, were removed and liquefied at gas plants and then poured and temporarily stored in acre-sized "blocks" nearby. When market conditions were favourable, the sulphur was re-liquefied, pelletized and then sent by rail to the ports on the west coast, and from there to commodity markets around the world. The technology was very effective, removing over 99 per cent of the sulphur from the raw gas.

As for the colourless, odourless and relatively benign CO_2, it was removed by a catalytic process and simply vented into the atmosphere. To my knowledge, no one in the company had calculated the volume of CO_2 the plants were releasing to atmosphere, as there had been no business reason to do so. Driven by my own curiosity and anticipating potential regulatory developments, I spent hundreds of hours sifting through reams of raw gas chemical composition data and crunched the numbers. Off the side of my desk, I determined that the company's facilities were collectively releasing well over 6,000 tonnes of CO_2 per day into the atmosphere. In fact it turned out that Westcoast Energy's gas plant in Fort Nelson was one of the largest point sources of CO_2 in North America, and the total annual facilities' emissions were the largest of any company operating in BC. I was the first to know, and immediately shared this information with my manager and the board of directors.[41]

Around the same time, and within this context, I participated in a climate change conference in Vancouver where I first met Dr. Patrick Moore. I knew of Patrick as a co-founder and former president of the original Greenpeace, which got its start in a Vancouver west side basement in 1971. Of course Greenpeace is a name now known to just about everyone in the developed world. Patrick's tenure with Greenpeace had taken him around the world many times and placed him in many challenging situations. In the mid-Pacific Ocean, he positioned himself precariously on a zodiac inflatable between frustrated Russian whalers and the huge mammals they were attempting to harpoon, kill and render for the many tonnes of valuable oil and flesh each carried. The actions taken by Patrick and other like-minded individuals in aid of respecting and protecting whales and seals have reverberated around the planet ever since.

These activities were covered by everyone from the mainstream media to *Penthouse* magazine and drew support from some surprising sources. For example, the CIA reportedly supplied the *Rainbow Warrior* with regular updates on the location of the Russian whaling fleet. And when the well-known personality Brigitte Bardot was drawn into the seal pup slaughter drama on the St. Lawrence River ice flows in Quebec, the media attention went global.

Revered by some, mistrusted by others, Patrick wasn't to be trifled with. Never without a whack of facts in his back pocket, Patrick was challenging in any debate. As his positions were most often science-based, I didn't often find myself at odds with him. Where he, in my opinion, has strayed from science in taking a policy position, I have made my position known, and we have agreed to disagree.

Patrick resigned as the Greenpeace president in 1986 after looking around the table at a board meeting and seeing there were no other scientists present. It was difficult if not impossible to garner support for science-based policies and actions without scientific bench strength. And there were new issues on the horizon.

THE CARBON PROJECT

In 1989 Patrick took a break from oceans, whales and media stunts to undertake a new initiative: "The BC Carbon Project." Supported through a modest grant that Patrick and the forest industry secured from the BC Science Council, the project gathered the interests of a group of open-minded and committed scientists, managers and engineers working in academe, industry and government, as well as a Greenpeace representative. The common denominator was an interest in understanding where BC's industrial CO_2 emissions were coming from, the quantities and what opportunities there were for reducing them.

Wearing my Westcoast Energy hat, I supported the Carbon Project by representing the upstream natural gas industry, openly supplying emissions data I had secured relating to raw gas production, processing and transmission. BC Gas, the province's primary natural gas retailer, was looking into emissions associated with gas distribution to retail customers. BC Hydro was investigating possible emissions associated with reservoirs and power generation, while MacMillan Bloedel Ltd. was looking at emissions from some of its BC forest operations.

The Carbon Project marked a first significant step in a much broader and better resourced range of initiatives undertaken by Canadian industries to get on top of the climate issue. Co-operating with and staying abreast of the environmental organizations, some of which were struggling with falling public interest in their respective causes and who saw climate as a renewed focus for fundraising, was an objective as well.

When I joined Westcoast Energy Inc. the company was an active member of the Canadian Gas Association (CGA). The CGA had established an Environment Committee to address a wide range of issues associated with the operation of gas facilities. Natural gas production and delivery are complex, involving raw gas wells, gathering systems, gas processing plants, long-distance transmission by pipeline to distributors and the local delivery of gas to industrial and domestic consumers. There was no shortage of issues to cover.

The membership represented Canada's largest companies in the natural gas industry, such as Consumers Gas, Union Gas, Centra Gas, TransCanada Pipelines and BC Gas, as well as companies that supplied equipment and services to the industry. For example Rolls Royce, a major manufacturer of jet turbines designed for gas transmission facilities, had to deal with the engineering challenges of emissions regulations, and thus was always represented. I was asked to represent Westcoast at the meetings and did so. Until this point in my life, I had only known about big companies from the outside looking in.

Anyone who has been on the inside of a large company knows that the appearance from the outside has been created in the public affairs or corporate communications department. While the picture presented is not fictitious, it is carefully managed and may fall short of a complete portrayal. Now I was "inside," and moreover, I was party to the goings-on of a national energy industry association, including its lobbying in Ottawa – again, from the inside.

This positioning, and the fact that my company was integral to the formation of another industry association, the Canadian Energy Pipelines Association (CEPA),[42] meant I was going to have an unprecedented look at the workings of the oil and gas industry in Canada from a greenhouse gas point of view.

I found to my delight that the CGA committee members were capable, committed and spirited. These were certainly not the "industry apologists" I was expecting. Nobody was ducking issues. We worked hard. We played hard. We sometimes referred to ourselves as the "gas heads."

The committee met quarterly to share environmental best practices, to receive updates on proposed government policies and environmental legislation from the CGA executive and to discuss environmental issues germane to the industry. The meetings were methodical and always to an agenda. I had no idea I was about to be party to a revolt.

REVOLUTION IN QUEBEC

It all began at a committee meeting in Quebec City,[43] in late 1994. At this meeting, the CGA's vice president of government affairs dutifully provided an overview of policy developments in Ottawa, including a synopsis of the federal government's activities around the emerging issue of "climate change," or "global warming" as it was more often referred to at that time. In his synopsis, the vice president commented that climate change "wasn't an issue" and would go away – indeed there was "nothing to be concerned about."

He then began to move to the next agenda item.

Being a newcomer and unaccustomed to national committee work, with some trepidation I interrupted the vice president and offered respectfully that I could not agree. I knew Canadian (and global) CO_2 emissions were growing in step with current energy development policies and markets, and I knew that Canada had played a lead role in the development and signing of the United Framework Convention on Climate Change (UNFCCC) at the Rio Summit in 1992. For signatories to the convention this meant that "climate mitigation," which refers to actions taken to reduce and or stabilize greenhouse gases (GHGs), and "climate adaptation," which focuses on preparing for and adapting to current or inevitable impacts of climate change, were both going to have to be addressed by large emitters of GHGs.

The spontaneous revolt began to gather momentum when Marika Hare, the director of environmental affairs for Consumers Gas, and someone who clearly understood the importance of the issue to her company and the gas industry as a whole, spoke up, as did Bob Bell, manager of environmental affairs for Union Gas and Centra Gas, who recalls:

> *We were the first wave of senior environment professionals (holding management positions) in the industry ... and it really was unanimous. We were truly a group of committed and concerned professionals who stood up to the old guard and said 'NO, this is important and it will continue to be so.'*

At first the startled CGA vice president was taken aback at the apparent audacity of the committee members, but he received the feedback professionally. And so began a fundamental shift in strategy from denial or dismissal to transparency and co-operation.

FEDERAL POLICY CONUNDRUM

Clearly, Canada was (and is) facing a policy conundrum. On the one hand, fossil fuel energy development policies and markets were driving CO_2 production and operations emissions relentlessly upward. On the other hand, in signing and ratifying the UNFCCC, Canada had agreed to a target of reducing emissions to 1990 levels by 2000. The country was committing itself to achieving the impossible. Canada's energy development and climate mitigation policies were on a collision course, and the situation surely needed attention – by government, by industry,[44] by virtually everyone.

Our modest Environment Committee could see it. But apparently, the CGA Board did not.

Not only had it missed this point, but the board had decided to diminish the role of our Environment Committee to a subcommittee of the CGA Operations Committee. Faced with the demotion, the renegade Environment Committee simply said "NO," went in the opposite direction and carried on.

Marika Hare recalls:

> *When told, the Environment Committee did not accept this news. We disagreed that the Committee should be a subcommittee of the Operations Committee. We argued that climate change, emissions measurement and mitigation, carbon trading were priority and strategic issues in our industry. Rather than fold, we simply planned our next meeting. We were going to carry on with the work that was started because we all understood its importance.*[45]

RESTORED PURPOSE

From this point forward, "climate change" was a standard item on the committee's quarterly meeting agenda. Also, a number of proactive actions were taken, including the establishment and endorsement by the CGA Board of Sustainability Guidelines and an Environmental Code of Practice, which I was honoured to have drafted. The committee also decided to have an independent body undertake a comprehensive industry-wide inventory of greenhouse emissions from the natural gas industry across Canada, and more. A key objective of this commitment was to understand greenhouse gas sources better and to develop reliable means to measure, monitor and manage these emissions.

The elevated importance of the climate issue to the industry was confirmed when a CGA Environment Award was established to recognize industry contributions in a number of areas, including greenhouse gas emissions management.

THE SDRI CLIMATE DESK AND GEMCO

Under another initiative, the CGA's Environment Committee established and resourced a "climate desk" for a graduate student at UBC's Sustainable Development Research Institute (SDRI). SDRI was a think tank founded and headed by Dr. John Robinson, with faculty and associates that included Mike Harcourt (former Vancouver mayor and BC premier) and Dr. John MacDonald (co-founder of MacDonald, Dettwiler and Associates), a number of academic researchers, and me. SDRI was one of the first academic institutions in Canada to take on climate change and adaptation seriously, with Director John Robinson holding a co-chair role on an IPCC committee.

The CGA Environment Committee was also instrumental in the establishment of GEMCO, the Greenhouse Emissions Management Consortium, a unique undertaking that took the management of the climate and carbon issues to an unprecedented level, executing the first international trade of carbon offsets between parties in the US and Canada. (See Chapter 5.)

I have no doubt the CGA Environment Committee's spontaneous revolution in Quebec was somehow mirrored across the country and other industry sectors, as a wave of well-educated, pragmatic and dedicated professionals came to take their place at the issue management tables.

STEPPING UP AND REACHING OUT

Although the UNFCCC was signed in 1992 and ratified in 1994, there was no indication the Canadian government was going to implement anything in the way of domestic policy and regulations to meet its UNFCCC obligations. To her credit, in November 1995, Deputy Prime Minister Sheila Copps told the truth: Canada was *not* going to meet its UNFCC commitments. At the same time, industry associations such as the CGA, CEPA and CAPP[46] were becoming increasingly proactive, co-operating with Natural Resources Canada (NRCan) and other federal agencies and non-governmental organizations to establish the Voluntary Carbon Registry (VCR).

The VCR challenged companies to voluntarily establish greenhouse gas emissions inventories, identify emissions reductions opportunities, implement those that were both pragmatic and effective and to engage in offset project development. These activities were posted on the VCR website for anyone to peruse, investigate and challenge as appropriate.

At least as important as the work within industry and government was the reaching out to green organizations. At that time Elizabeth May, current leader of the federal Green Party, was executive director of the Sierra Club. Ms. May had (and has) a very strong interest in climate change and was not shy about expressing her opinions.

On the heels of the deputy prime minister's announcement that Canada would not meet its UNFCCC obligations in late February 1995, I agreed to discuss and debate the roles and responsibilities of the energy industry, vis-à-vis climate change, with Elizabeth May on CBC Radio's *Morningside* program, with the iconic (the late) Peter Gzowski as host. Ms. May directed the majority of her commentary at the apparent inaction of the federal government in implementing the policies of its own "Red Book."[47] However, she also implied that in the absence of government action the fossil fuel energy industry was dragging its heels and, worse, was uninterested in investing in emissions reduction opportunities such as energy efficiency and renewable energy as they were counter to industry interests.

To this Peter Gzowski asked me straight up, "So why no action, Robert?" The simple answer, that of the "policy conundrum," I didn't have the presence of mind to articulate. But in fact, investments in energy efficiency and renewables were well underway, and I invited Elizabeth to watch closely, "because if they[48] fail it's going to be pretty obvious." History has confirmed these investments, which have been in the hundreds of millions of dollars.

Beyond our ongoing and growing efforts to take action to identify and employ emissions reduction measures internally, and to develop capacity to undertake large-scale offset projects, we continued to reach out to more key environmental organizations and said, "Let's work together on this."

A REMARKABLE CHARACTER

Most notably at the climate table at this time, participating in virtually all the climate-related conferences across Canada, was the Pembina Institute. Led by its founder, the erudite Rob Macintosh,[49] an ex-teacher from Alberta's Drayton Valley, Pembina was very much "in the trenches." Rob started Pembina in 1982 as a one-time-only, single-purpose citizens' group, after a sour gas blowout occurred down the road from his Drayton Valley home and killed two workers.

Rob was always articulate and respectful in his climate presentations but pulled no punches. Surrounded by "suits," Rob was entirely at ease talking to any person at any level. His "suit" was a pair of jeans and a plaid shirt. Rob was trim, sporting waist-length red hair, often in a ponytail, and an equally vibrant beard. He was hard to miss.

I had seen Rob present several times to large audiences, using hand-drawn overhead transparencies prepared minutes before his presentation. We finally talked when I literally bumped into him in an airport coffee shop. I sensed an openness and started chatting with him as we filled our cups with burnt java. I invited him to join me and some of my industry colleagues.

Rob was open-minded but understandably saw me as an industry guy. He couldn't have known that in 1976 I too had co-founded an environmental group. The Telkwa Foundation was originally established to help local residents challenge an oil pipeline proposed to stretch from Kitimat to Chicago. Supertankers carrying Prudhoe Bay oil would be navigating through Douglas Channel to off-load crude in Kitimat. Whether or not the Telkwa Foundation had any influence, the pipeline proposal was abandoned. My tenure with the foundation was short, but it gave me a taste for advocacy, and I respected Rob's approach to it.

We comfortably started sharing our knowledge, with an underlying sense of co-operation and respect, and perhaps some urgency. Thereafter, whenever there was an opportunity, we would pull each other under our respective tents, to compare notes and share perspectives. The energy industry was terribly opaque. I am not sure if that was by design or a reflection of old-school naïveté. In any case, it served no good purpose, and the gas heads decided to let it all hang out: The good, the bad and the ugly.

The reality is that many boardrooms of giant energy companies had opened their doors and welcomed Rob's perspectives.

I had seen the energy industry from a number of angles, not all pretty, but now I was seeing it from a place of co-operation and transparency. I liked this place.

Rob exited the climate space in a remarkable fashion, leaving Pembina as a staffer in 2001, a few years after he stepped down as the organization's executive director. At the party in Calgary to honour Rob and wish him well were most, if not all, of the CEOs of big energy in Alberta. Rob was toasted and roasted. The hair was mentioned. A hat made its way around the tables, and by the end of its tour was apparently graced with close to $25,000 to support Pembina's work, as a thank-you to Rob for his honesty, determination and fair-mindedness. Rob made good on his sacrifice – the red hair came off.

SYNCHRONICITY IN THE ROCKIES

One of the last times I saw Rob was at the Columbia Icefields south of Jasper. He was one of two figures descending the face of a glacier. I was oblivious, looking incredulously at a peg in the bare ground that said "1974," not believing. If true, the glacier appeared to have retreated over a hundred metres since the peg was put in the ground, which happened to be the year I had last been at the icefields, riding my Triumph 650 Tiger motorcycle through a June snowstorm.

Climate change. Right here. Right now. Right in my face.

It was then, when I had fully grasped the climate reality that was before my eyes, and its pace, that I sensed a presence. I looked up to see Canada's senior climate pioneer activist guru, Rob Macintosh.

Standing a few feet away with bright eyes and a smile, he extended his hand one more time. "Hey Robert, I'd like you to meet my daughter..."

It was an extraordinary moment, the script from a perfect movie. Rob Macintosh, his daughter, the retreating glacier, and myself – intersecting as if by design. All captured in a timeless instant of absolute clarity around the interconnectedness of people, purpose and time.

The Greenhouse Emissions Management Consortium (GEMCO)

1994–1997

As the issue of climate change continues to gather momentum, the mechanism known as "carbon trading" will become increasingly relevant. And so, before telling the story of how the Greenhouse Emissions Management Consortium (GEMCO) came to be one of the first and largest industrial consortia of carbon offset trading participants in the world, a brief backgrounder on "carbon offsets" and "carbon credits" is warranted.

WHAT IS A "CARBON OFFSET"?

As it can neither be seen, touched nor heard, it should come as no surprise that the term "carbon offset" is poorly understood. However, given that global carbon trading has exceeded US$100-billion per annum, it is fair to say that carbon offsets, while intangible, are arguably real and definitely have value.

The World Resources Institute (WRI) defines a carbon offset as "a unit of carbon dioxide-equivalent (CO_2e) that is reduced, avoided, or sequestered (from the atmosphere) to compensate for emissions occurring elsewhere."

As a functional, tradable unit, carbon offsets are measured in metric tonnes of carbon dioxide equivalents (abbreviated to "CO_2e"). The CO_2e term was established to accommodate other greenhouses gases used in carbon offset trading such as methane (CH_4), nitrous oxide (N_2O), chlorofluorocarbons (CFCs), and others. As these other gases have different global warming potentials, they are all converted to one measure, or currency: tonnes of CO_2e.

A "carbon credit" is a carbon offset that meets a specific standard that qualifies it to be used by industrial emitters to address governmental emissions reductions policies, regulations and requirements, and/or to be fungible (tradable) in an emission reduction trading system, in an identified jurisdiction.

CARBON OFFSET ATTRIBUTES AND QUALIFIERS

The WRI definition is accurate but not adequate in conveying the full picture of what makes for a legitimate carbon offset. Before a carbon offset is considered genuine and tradable,[50] it must arise from a project meeting the following minimum requirements:

- *Real*: The project actually prevents the release of GHGS into the atmosphere, or sequesters CO_2 or the equivalent from the atmosphere.

- *Additional*: The project delivering emission reductions (creating the offsets) must go above and beyond "business as usual." If an action is planned in the future, or likely to happen anyway, it would not be considered "additional."
- *Verifiable/Quantifiable*: A project's carbon benefits must be quantifiable in order to determine the volume of carbon offsets being generated. It also must be independently verified under an accepted protocol[51] by a certified third party auditor, to ensure the legitimacy of the project and resulting offsets being produced. Once verified, offsets are serialized and posted on an internationally recognized registry.
- *Permanent*: The project must produce emission reductions that are permanent. For example, if a project captures CO_2 and pumps it into a deep geological formation, it must be ensured that the buried CO_2 is secure and does not leak back into the atmosphere. In the case of an afforestation project and the storage of carbon in forests, if a wildfire releases the stored carbon to the atmosphere, there must be a mechanism to replace any issued credits with an equivalent number of additional credits.

Carbon offset projects take on many forms.

Examples include projects such as sustainable agriculture, avoided deforestation, improved forest management and forest ecosystem restoration. Such projects capture and store carbon in living biomass (trees) and in the soil.

Other project types are those of a technological nature.

Examples include renewable energy systems (biomass, wind, tidal and solar); technology transfers; fuel switching; and energy efficiency projects.

Specific, real-life examples of carbon offset projects developed by GEMCo are presented later in this chapter.

WHO BUYS CARBON OFFSETS?

There are two markets for carbon offsets. The "Voluntary Carbon Market" was the first to appear, evolving in response to the growing population of individuals wanting to offset their household or travel CO_2 footprint voluntarily because they are committed to lessening their net impact on the environment.

The Voluntary Market also addresses the needs of businesses and other organizations that for policy reasons or for promotional purposes require their operations and/or products to be branded as being *carbon neutral, climate friendly* or some variant thereof in order to attract ethical investors and/or customers, and secure a "social licence to operate" from the public.

While the Voluntary Market is relatively small, it goes through periods of growth. Findings published by Ecosystem Marketplace in May 2016 showed an estimated 84 million tonnes of carbon offsets transacted in 2015 at a total value of US$278-million, with a trend of increasing volumes and lower prices.[52]

The second type of market to emerge, the "Compliance Carbon Markets," are driven by government regulations directed to net emissions reductions within a jurisdiction (whether at an international, country, state, provincial or regional level). These programs accommodate the trading of carbon offsets among emitters subject to the regulation. The largest compliance market is the European Union's Emission Trading System (EU ETS), which became operational in 2005 and was the first cap and trade scheme for carbon emissions. The EU ETS includes the participation of approximately 12,000 reporting stationary installations and 1,300 reporting aircraft operators.

The EU ETS includes Austria, Belgium, Bulgaria, Croatia, Cyprus, Czech Republic, Denmark, Estonia, Finland, France, Germany, Greece, Hungary, Iceland, Ireland, Italy, Latvia, Liechtenstein, Lithuania, Luxembourg, Malta, Netherlands, Norway, Poland, Portugal, Romania, Slovakia, Slovenia, Spain, Sweden and the United Kingdom. Annual trading volumes now exceed 8 billion tonnes of CO_2e per annum.

Other jurisdictions and regions have established trading systems or are in the process of doing so. In North America, the Province of Alberta was the first jurisdiction to price carbon and mandate emission reductions. Alberta was also the first jurisdiction in North America to utilize offsets as a mechanism to help emitters comply with provincial regulations. The Regional Greenhouse Gas Initiative (RGGI) in the northeastern United States was the next emissions trading system to develop and the first involving multiple jurisdictions. It includes industrial emitters in nine US states – Connecticut, Delaware, Maine, Maryland, Massachusetts, New Hampshire, New York, Rhode Island and Vermont.

In January 2013, California launched a full cap and trade system[53] known as "AB 32" (Assembly Bill 32) that mandates GHG emissions reductions out to 2030 and beyond. California has a rather robust carbon offset system. Shortly after California launched, the Province of Quebec passed its own cap and trade policy, which is aligning with California's. The Province of Ontario launched its cap and trade system on January 1, 2017. Other jurisdictions already have markets, such as the Province of British Columbia which has a well established carbon neutral government policy, as well as regulations to address liquefied natural gas (LNG) operations. Manitoba is also in the process of developing its own trading systems, anticipating eventual alignment with other jurisdictions.

Will these systems formally link with each other at some future date? Certainly, such "harmonization" would deliver efficiencies, and informal talks are underway among trading systems. However, climate policy has a history of proceeding at a very slow pace, so the development of a national or multinational system in North America is likely to remain at the informal discussion level for the foreseeable future.

WHY EMISSIONS TRADING?

The primary reason for using an emission trading system is that, if designed properly, it will identify and resource the least-cost solutions in meeting a shared emission reduction objective. In the case of a trading system, all parties share the reduced cost of meeting the regulatory objective. The success in addressing acid rain in the US was largely achieved through the implementation of an emission trading system.

There are no "losers" in implementing an emission trading system, except perhaps the governments seeking additional revenues in the alternative case of a carbon tax, and the bureaucracies that would be required to establish, resource and enforce alternative "command and control" agencies and regulations. For example, a regulation that requires all subject emitters to reduce CO_2 emissions by a fixed percentage, say 10 per cent, will incur higher net costs of achieving the reductions. While such a regulation may aim to achieve the same environmental benefit as a trading system, it requires the 10 per cent emission reductions to be found within every emitter's facility, regardless of costs to each, which will be highly variable. Furthermore, there will be additional costs associated with ensuring compliance with the regulation.

In summary, by giving emitters access to less expensive but equally effective emission reduction projects outside their fence-lines, a trading scheme can significantly reduce the overall costs, for both the emitter (business) and the regulator (a government agency), in achieving the regulated emissions reduction objective.[54]

At an international scale, the cost savings of meeting emission reduction goals using a well-designed emission trading system will be measured in the billions of dollars per year, and it will do so without compromising the environmental objectives.

Another advantage of emission trading is that it gives rise to new commerce from incremental science-based businesses that offer, for example, greenhouse gas accounting software products, as well as validation, certification and registration services. In addition, emerging low-carbon industries and clean-tech businesses, like wind, tidal and solar energy and energy efficiency software developers, will be supported.

Designed and implemented properly, carbon offset trading works. Arguably, the carbon offset industry had a very bumpy beginning, with no standards, no rules, no accredited organizations and some fly-by-night opportunists looking for a quick buck. Like the "wild west" of the 1800s, those days are gone, and just about anyone in the industry would likely argue that it has gone to an extreme in the other direction, extraordinary rigour.[55]

THE PRAIRIE SOILS PROJECT AND THE GENESIS OF GEMCO

In the spring of 1994 I was invited by TransAlta Corporation to a presentation in Calgary to hear about a large carbon offset project proposal. At the meeting I learned that TransAlta, a recognized leader in sustainability policy, had issued a Request for Carbon Offset Proposals early in 1994. After assessing a host of project proposals, the TransAlta Sustainable Development team chose to focus on the "Prairie Soils Carbon Balance Project."

The project proposal was directed to making significant changes in various agricultural practices, and these were expected to produce significant volumes of low-cost carbon offsets.

I was intrigued by the Prairie Soils Carbon Balance Project and agreed to attend a subsequent meeting, this time at TransAlta's offices in Edmonton. The group had shrunk, but I was not alone. The environment and sustainability managers and directors from several Canadian energy companies crammed into a small room at TransAlta's offices in Edmonton to hear details of the proposed project. The meeting was called by Dawn Farrell. Dawn is currently

President and Chief Executive Officer of TransAlta Corporation; at that time Dawn was TransAlta's sustainable development director.

Dawn wished to provide more details on the prairie soils carbon sequestration project to a focus group of like-minded individuals who were carrying the sustainability flags for their respective companies, and was hoping to identify some industry partners to participate in the project.

TAKING THE LID OFF

The Prairie Soils opportunity showed significant potential, but as I listened, I recognized that there were many other projects being proposed, by a wide range of parties, and that many more projects would need to be implemented if there was to be a meaningful impact on net emissions. With this in mind, at some point in the meeting I found myself stepping to the whiteboard, suggesting we "take the lid off" and focus not just on prairie soils sequestration. I suggested we establish a greenhouse emissions management consortium that would advance a *portfolio* of offset projects. The words were hardly out of my mouth and on the whiteboard before me when a voice from the back of the room, belonging to Perry Toms,[56] a very sharp mind with Dawn's sustainability team at TransAlta Corporation, chimed in from the back of the small room with "GEMCo."

Thus the unique GEMCo vehicle was conceived. No one at that point could know that GEMCo would become a pioneer and international leader for developing contract and other legal language, supporting soils sequestration and other science, developing and managing a range of projects, and be the first to facilitate a large multi-party international carbon trade.

As the leader of Westcoast's newly established Sustainable Development Office and Council, I had recently hired Aldyen Donnelly to be our senior economist. Given her background, GEMCo seemed a good fit for Aldyen, and so I handed her the GEMCo file to manage. Over the next year Aldyen ran with it.

The consortium contracted with Murray Ward,[57] an international carbon policy expert now operating from Vancouver, to write the business plan. The plan's introduction, dated October 2, 2005, read as follows:

> *The Greenhouse Emissions Management Consortium (GEMCo) is a proposed not-for-profit corporation formed by Canadian companies to demonstrate industry leadership in developing voluntary and market-based approaches to greenhouse gas management. The effort to limit these emissions has emerged as a major political and environmental challenge for the Canadian government and a pressing economic factor for Canadian Industry.*

To no one's surprise, when it came time to hire a GEMCo president, Aldyen threw her hat in the ring for the president's job at the eleventh hour. The membership chose her to lead the consortium, and she left Westcoast to operationalize GEMCo.

Aldyen was a tour de force in the carbon world. Always controversial, never without an opinion, endowed with a steel trap mind for data, boundless energy and a very big heart, Aldyen made things happen.

CHAIR TODAY, GONE TO MAUI ...

As it turned out, in some strange irony, I was departing from Westcoast when GEMCo finally incorporated and became operational. As I would no longer be associated with any of the member companies, my role as chairman of the board of the newly incorporated not-for-profit vehicle lasted for the inaugural board meeting in Calgary, then immediately ended.[58]

However, the structure I had fought for was established. Aldyen was at the helm, and consistent with its original intentions, GEMCo formally launched the Canadian Prairie Soils Carbon Balance Project. This project represented a research partnership with Agriculture and Agri-Food Canada, Alberta Agriculture, Food and Rural Development, Saskatchewan Soils Conservation Association, Ducks Unlimited and the Canadian Cattlemen's Association.

The Canadian Prairie Soils Carbon Balance Project was originally focused on the measurement and collection of soil carbon data through to the year 2000 at approximately 200 locations across the Canadian prairies where farmers had adopted sustainable land management practices, such as continuous cropping and (no till) direct seeding. The research program involved analysis of new data as well as a large pool of historic soil carbon information. Activities included the development of a sampling protocol that would allow the verification of carbon changes arising from a range of best management practices for annual cropping and for grassland production. A number of soils scientists were engaged to ensure rigour was applied to the calculation and measurement of carbon benefits.

AN END TO "PIE-IN-THE-SKY"

GEMCo also worked with carbon-savvy law firms to establish capacity in terms of contract language to facilitate carbon trades, and executed the very first cross-border trade of carbon between farmers in Iowa and GEMCo members in Canada. Carlton Bartels, then director of emissions trading for Cantor Fitzgerald and CEO of CO_2E.com, the trading organizations that brokered the deal, commented, "this Agreement is a bit of a trailblazer for North America."

Other projects followed.

On October 19, 1999, GEMCo announced an agreement with IGF Insurance Company, a large crop insurer in the US, to buy up to 2.8 million tonnes of carbon credits from the "Iowa Farm Project." Seven consortium members participated in the agreement, which was to run through 2012.

The agreement was the first of its kind in applying a range of agricultural sources to generate carbon credits. Until this transaction, "the talk about carbon credits had been pretty much pie-in-the-sky," said Steve Griffin, IGF vice president for strategic development at that time.

Another project followed. GEMCo's "Norseman Project," announced in November 2000, was founded on methane emission reductions accomplished through burner fuel modifications in a wallboard plant in Surrey, BC. The emission reductions occurred when methane that would have otherwise been released into the atmosphere from the landfill was used to fuel the process, and was thereby converted to carbon dioxide.[59] In its agreement with GEMCo, Norseman agreed to deliver up to 301,000 Emission Reduction Credits (ERCs) over a 14-year period.

In November 2000, GEMCo also announced the "Petro Source Project," which involved the capture of CO_2 that would otherwise be vented to the atmosphere from flue stacks at a number of natural gas treatment facilities, using the Val Verde CO_2 pipeline in Texas. The captured CO_2 was to be used for enhanced oil recovery, with the CO_2 being sequestered in geological formations in the process. Six GEMCo members were to participate in the option agreement delivering up to 600,000 tonnes of carbon offset credits from 2002 to 2012.

Another GEMCo project, announced in September 2004, was the "Integrated Gas Recovery Systems Project," located in Niagara Falls, Ontario. This undertaking involved the collection, compression and delivery of methane-containing landfill gas that would normally be released to the atmosphere. Instead the gas was to be transported three kilometres by pipeline to a paper mill, blended with natural gas and used to fuel the boilers. GEMCo announced it would transact under a firm forward agreement that required the company to reduce GHG emissions at their Ontario operations by a total of 850,000 tonnes of CO_2e over a ten-year term.

GEMCo's membership was impressive, growing to 11 corporate members that directly employed over 60,000 Canadians and owned over $60-billion in energy assets. Collectively, these companies provided a range of energy services to over 15 million Canadian customers. The GEMCo membership represented roughly 25 per cent of Canada's industrial greenhouse gas emissions.

MISSION ACCOMPLISHED

Having blazed a trail, having ended the "pie in the sky" era, having yielded a number of "firsts" in the carbon trading industry, and having kick-started international carbon offset trades, GEMCo's original *raison d'être* was largely addressed. With the growing wave of capacity development in validation protocols, verification standards and carbon registries arising from a range of sources around the world, GEMCo began to step back from the limelight, its original mission as a catalyst accomplished.

From right to left: Dr. Wangari Maathai, Dan Rather, and Al Gore at the Yale Club, NYC, September 22, 2008.
PHOTO CREDIT: JEFF HOROWITZ

The pair of white cowboy boots from the Dawson City dance that sparked the international voluntary forest carbon offset market.
PHOTO CREDIT: ERIN KENDALL

White-sided dolphins cavort in Howe Sound.
PHOTO CREDIT: JESSICA HAYDAHL RICHARDSON

Catch, measure, and release: a healthy juvenile steelhead trout in a Fraser Valley river.
PHOTO CREDIT: ROBERT W. FALLS

A bald eagle maneuvers towards a soon-to-be airborne salmon resting near the surface waters of Howe Sound.
PHOTO CREDIT: JESSICA HAYDAHL RICHARDSON

A Pacific humpback whale about to sound in the Salish Sea.
PHOTO CREDIT: JESSICA HAYDAHL RICHARDSON

A friendly harbour seal in Fisherman's Cove, Howe Sound.
PHOTO CREDIT: ROBERT W. FALLS

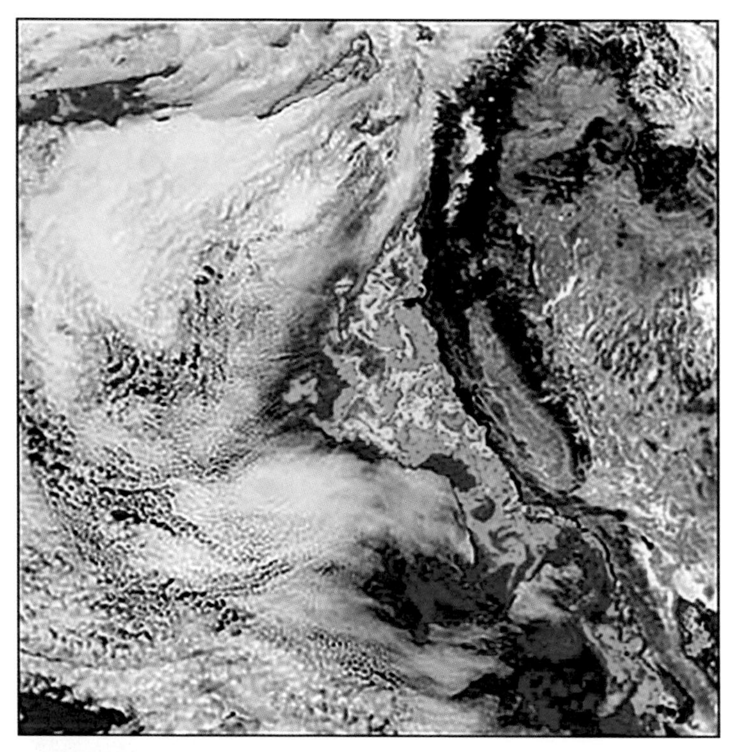

A SeaWiFS image of phytoplankton chlorophyll concentrations in the California Current System.
PHOTO CREDIT: NASA SEAWIFS PROJECT, GODDARD FLIGHT CENTER AND GEOEYE.

A Bell Jet Ranger helicopter supporting aquatic inventories on the Petitot River in northeast B.C.
PHOTO CREDIT: ROBERT W. FALLS

Rugged, mineral-rich coast mountains along the northeast corner of the Salish Sea.
PHOTO CREDIT: JESSICA HAYDAHL RICHARDSON

A local paddle boarder enjoys the company of Pacific white-sided dolphins in Howe Sound.
PHOTO CREDIT:
JESSICA HAYDAHL RICHARDSON

An Orca whale travelling to a feeding ground on the Salish Sea. In this area, Orcas feed on salmon, seals, sea lions, porpoises, and dolphins.
PHOTO CREDIT:
JESSICA HAYDAHL RICHARDSON

The author at his home on the Salish Sea.

China: Technology Transfer & the Buddha's Finger

1997–2003

I continued my work with the Canadian gas industry until late 1996, when during a re-organization I was presented with the opportunity to either stay on as leader of the Sustainable Development Office at Westcoast Energy or leave to carry on with my career from another stage.

The decision was quick. I saw this as a once in a lifetime chance. I took the plunge and left Westcoast Energy to focus on renewable energy, ecosystem restoration and other types of climate mitigation as my primary foci, as an entrepreneur. I reaffirmed my connections at UBC in order to sustain the mutually beneficial bridges we had built between academe and industry.

My career with Westcoast Energy ended formally at a dinner celebration in Windsor, Ontario, after a Sustainable Development

Council meeting which was held at Union Energy headquarters in nearby Chatham. Following the dinner party, a number of council members headed across the river to Detroit for some less formal celebrations. It was an evening made memorable by extraordinary music and an early morning return to Windsor in a taxi occupied by a significant population of brazen cockroaches – and the driver. The vehicle was home to both, but it got us back to Canada.

BANFF SCHOOL OF MANAGEMENT

A month or so later, as I continued to plot my plan forward, I was contacted by my friend and associate John Woodruffe, who invited me to be a guest faculty member at the Banff School of Management. John was teaching a course on environmental and sustainability reporting and needed a qualified instructor, as his fellow faculty member had taken ill.

This opportunity was somewhat predicated upon our report writing success at Westcoast Energy. The company's first progress report on implementing sustainable development netted a Silver Award from the *Financial Post* in 1994. John had prepared sustainability reports for Shell Canada and a number of other large Canadian energy companies. His fastidious attention to good writing, detail, accuracy and the picayune was at times frustrating but resulted in a fine product. John was a pleasure to work with and we made a great team.

So I said yes, and a week or so later found myself sipping a chilled lager with John at the student lounge in the Banff School of Management the evening before we were to begin the course.

John knew my growing penchant for all things carbon and posed a question – the answer to which would have us both very busy for the next several years, preparing to launch one of China's first climate change projects during the first term of the Kyoto Protocol.

John had worked with Shell's Sulphur Marketing Team and continued to consult with them, helping negotiate multi-million-dollar transactions in elemental sulphur, a by-product of gas and oil production, around the world. Consequently he was party to the company's market development activities. There had been recent market strategy work done by the team directed to dramatically growing sales of sulphur to China. With a huge population to feed, China had a very large appetite for sulphur to produce sulphuric acid, a key component in fertilizer production.

The marketing team had identified an opportunity to supply elemental sulphur to sulphuric acid producers that were currently using pyrite as the feedstock. Pyrite is comprised of sulphur, carbon, mercury and a potpourri of unsavoury and toxic chemicals and elements. When combusted by roasting with coal, the pyrite's sulphur was released from the rock and captured and used to produce acid. The rest of the pyrite's toxic contents and coal ash went back into the environment in a range of waste streams. The release of these contents was polluting the air, poisoning groundwater and clogging overflowing landfills with solid waste.

Since carbon was one of the constituents of the pyrite, upon combustion CO_2 was produced and released to the atmosphere. Given the huge number of sulphuric acid plants consuming pyrite, the total CO_2 released to the atmosphere numbered in the millions of tonnes.

INDUSTRIAL ECOLOGY

Fundamental to the opportunity was the fact that Shell had identified a patented technology for converting these sulphuric acid plants to burn pure elemental sulphur. By doing so, virtually all of the pollution, including the CO_2 arising from the combusted carbon, would be eliminated. What had been considered a waste product from Canada's gas fields would now be in demand as a highly valued commodity and the feedstock for the clean production of sulphuric acid, making the project an elegant demonstration of industrial ecology.

John's question to me in the Banff lounge was this: Could such a project as this qualify as a "carbon offset" project and could the arising carbon offsets be used to mitigate carbon emissions back in Canada? At this time, such a project could be considered within the context of the "Joint Implementation" (JI) provision under the UNFCCC rules. My answer was, "It looks promising, let's find out."

We pitched a feasibility study to Shell Canada management. After a pregnant pause, the senior executive responsible for environmental matters, who happened to be a medical doctor, commented that their group would "never have thought of this." More importantly, they gave us the green light to develop the proposal.

We were in business. We formed a two-man-band company, named it "International Offsets Unlimited Inc." and got to work.

SOJOURN TO OTTAWA AND WASHINGTON, DC

In order to explore the policy context, we flew to Ottawa to meet the Canadian government's Joint Implementation (JI) expert. The gentleman's name escapes me, but I recall finding him working alone in an open office overlooking a lush green space. Looking around, it was clear that JI was not his primary focus; in fact he had barely heard of it. Nevertheless, there is no doubt that some memo filed and collecting dust somewhere in the Byzantine environment stated the he was responsible for Canada's JI programming – such as it was. The man's honesty was refreshing, and appreciated. Cutting to the chase, he simply suggested that we "go to Washington" to meet some folks who were truly "on it." He provided us some names and off we went to continue our sojourn on Capitol Hill.

The United States' JI office security was intimidating. After being searched for weapons, we were led to a large, open work space where a dozen or so folks were toiling away on JI projects. They were pumped. They were serious. They were funded. And they were very helpful, suggesting, upon hearing of our proposal, that we just "go for it."

We also met a Canadian consultant to the UN residing in Georgetown, Sid Embree. Sid was and is a very bright woman, business-oriented, who was right on top of her game. We learned a great deal as a consequence of our man in Ottawa sending us to these elevated sources of JI intelligence in Washington. We had what we needed, and returned to our respective homes.

On the basis of these and other findings, and a thorough technical review of the proposed China project, we returned to Shell with a recommendation to proceed.

Shell agreed and the game was on.

NEXT STOP BEIJING

Beginning in January 1997, we began a series of visits to China to get the technology transfer project moving. One trip coincided with the COP 3 in Kyoto, Japan. Arising from that COP (Conference of the Parties) was the Kyoto Protocol, and attached to that was the Clean Development Mechanism (CDM). CDM promoted "clean" carbon mitigation projects in developing countries, sponsored by developed country companies who would be able to apply the carbon benefits in their home jurisdiction.

The CDM announcement was very timely and fit our intentions perfectly. We recast our proposal within the new policy context and began developing relationships with the appropriate authorities in China and key Canadian embassy operatives.

To our good fortune, during one of our trips to Beijing, Canada and China signed a CDM Cooperation Agreement that had Canada assisting China in getting some projects off the ground. The signatory for Canada, Minister Stéphane Dion, was carbon savvy, and our project appeared to fit within the UNFCCC's Clean Development Mechanism (CDM) perfectly.

This work in China was facilitated by a very capable and connected Shell agent who had grown up in Beijing a few blocks from Tiananmen Square. In fact the agent had heard the 1989 disaster unfolding and showed us where he huddled in his nearby home at the time, which was still occupied by his mother and other relatives.

Many of our contacts were located in Beijing, and the agent would set up the meetings with senior people in the national ministries of Energy, Environment and Finance. The challenge was to navigate to a range of venues in and around Beijing through traffic jams that were barely believable. In one case, I recall us driving up the sidewalk on the wrong side of the street, within a sea of people and bicycles.

I recall that no one honked during the traffic jams, and there was virtually no eye contact among the drivers. Cars, trucks and motorcycles positioning and jockeying within a centimetre of each other was the rule rather than the exception. I never saw an accident.

Another outstanding aspect of the city was how fast it was changing. Where there was an empty urban block of land one month, there would be a community forest ten metres high on returning six months later. Buildings went up seemingly overnight. Extraordinary architecture abounded. The new hotels we saw were ultra-modern and extremely high-end in all aspects. There were energy efficiency systems in the guest rooms I had not yet seen in North America or Europe.

And as if to make us Canadians feel welcome, the underground mall that ran from the Kerry Hotel, where we stayed, to the China World Hotel a kilometre or two away sported an ice hockey rink.

The human price paid for this extraordinary pace of building is a matter of conjecture. I recall looking out my window late one night to see a worker perched ten floors up on some new building, welding in the dark and bitter cold. I could see no helmet, and I believe he was wearing running shoes.

A STIRRING DRAGON

For the most part, the people we met were courteous and proud of their country. We learned that the Chinese culture planned in thousand-year time frames. The young people I spoke with felt the Dragon was stirring, and they were correct. China's economy was about to accelerate dramatically. Not surprisingly, so were its carbon emissions. During our first visit, China's CO_2 emissions were just over 3 billion tonnes per annum. The last reported measures showed China's emissions had grown to over 10 billion tonnes, representing 29 per cent of global emissions.[60]

Our mission required meeting regional officials as well, and of course we needed to see the target conversion plants. So we flew to Guiyang, Xi'an and other cities, and then took to the roads and highways, experiencing even more driving chaos. On one occasion, as we drove toward the city of Guiyang after touring a very large, modern sulphuric acid facility, our van driver decided to pass the bus ahead of us. We were on a two-lane highway having no shoulder, with a hilltop directly in front of us. As he pulled out, sure enough a truck emerged over the hilltop and was coming our way. I simply closed my eyes. I don't know how the collision was avoided, but given the fact I am currently breathing and typing, it apparently was.

Almost as frightening was participating in the Chinese approach to business entertainment. On one occasion, I recall being beckoned onto a stage by a hypnotic Caucasian-Chinese woman – of course there was no saying "no" – in order to have me participate in the highly popular "snake dance." Somewhere in the dance the "splits" were required, and I took a pre-emptive bailout from the stage to return to a table of howling Canadian and Chinese business partners.

Freeways appeared as quickly as buildings. We watched in disbelief as labourers climbed multiple wobbly wooden ladders, sometimes three atop each other, to hand-place giant boulders on the banks of the new highways – this in the brutal cold of winter, with minimal clothing and virtually no protective gear.

TOMBS EVERYWHERE

There was always a new experience. Flying in to Xi'an I saw a number of rectangular black blocks, perhaps a few acres each in size. I assumed that I was seeing stored coal. Wrong. These were in fact tombs. And there were hundreds of them. Of course: Xi'an was the home of the famous Terracotta Army. This archeological site was a true marvel.

A farmer had discovered the buried terracotta soldiers, mounted on their horses and numbering in thousands, while digging for a new well in 1974. The site, once quiet farmland, was now covered in a football field-sized enclosure, and the soldiers and horses were still being revealed with soft brushes and painstakingly excavated. Thousands had been recovered, untold thousands remained. Remarkably, the farmer who discovered the site was still there, busily autographing books about the discovery.

THE BUDDHA'S FINGER

During another road trip, we took a diversion to see a temple where on display was the Buddha's finger. After a short wait, we made our donations and entered the temple and began to peruse a collection of artifacts. However, the only item I recall with absolute clarity now is the Buddha's finger. The story held that when the Buddha died, his body was dismembered and pieces sent around the world to spread his influence. A finger had come to this temple and had been enshrined there ever since.

I made my way past a number of showcases, and finally there it was: the finger. Well, a piece of bone at least. Taxonomy is not my strong suit, and I had trouble imagining the bone I saw having been on the Buddha, or any human for that matter. It seemed far too large, and the shape was wrong, I thought. But then I had never seen a human finger on display. Given the minimalist security, my guess is that this was one of a lineage of "Buddha's finger" facsimiles, but that is not likely to ever be confirmed.

There were more surprises. One particularly grand tomb had a grand causeway leading to it. On either side were huge stone carvings, depicting the heads and shoulders of local and distant leaders. I wandered off the causeway and over the side of a hill to take in the scene of an orchard and a small community, perhaps of a dozen people. They were living in multiple-room cave homes that had been carved out of the hillside, and fashioning souvenirs, often small rock carvings, which they sold to the tourists nearby.

NON-STOP MEETINGS

Getting back to the business of carbon, our virtually non-stop meetings with officials took on a familiar pattern. Following a chaotic drive through the city, we would arrive at a building, enter, usually as a team of four, and begin looking for our meeting site. The interiors of the buildings were almost always dark, with lighting usually consisting of one bare fluorescent bulb per room. And if it was winter, they were cold. Ultimately we would be met and taken to the meeting room. In contrast, the meeting rooms had the feel and look of boardrooms back home, except they were grander. And thankfully, they were warm. After we entered and were graciously greeted, tea and perhaps an apple or orange would be served. The senior officials usually joined after we were seated. They were impeccably dressed in suits and were clearly at senior levels within the government. Most times serious, sometimes playful, they were always gracious.

The senior officials politely listened as we spoke of our project through a translator. Most often we would make our presentation, then there would be questions, and then a dialogue around our proposed Clean Development Mechanism (CDM) project would ensue. This usually resulted in much muted discussion among the officials. Thereafter there would be some pleasantries exchanged, an agreement to meet again would be secured, then off we would go to the next meeting.

From 1997 to 2002, we made a half dozen or more visits to China. The technology transfer portion of the project was proceeding nicely. Ultimately over 50 sulphuric acid plants were converted from pyrite-burning plants, with the associated pollution, to clean sulphur-burning plants. As a consequence, huge volumes of toxic wastes were eliminated, and millions of tonnes of CO_2 reductions were achieved.

MISSION ACCOMPLISHED – *ALMOST*

The process of diplomacy did not move quickly for us. Our final visit, which occurred in late 2002, had us meeting with senior Chinese officials and diplomats at the Canadian embassy in Beijing. The Severe Acute Respiratory Syndrome (SARS) scare was gathering momentum at that time, dominating the press. The Chinese decided to focus on the SARS problem, and our proposed CDM project, initially poised to be possibly the first CDM project in China, was placed on a back burner. This meant that there would be no formal "credit" claimed regarding the emissions reductions, beyond Shell identifying the carbon and co-benefits in an annual report. I expect Shell could have fought successfully to get the project back on the CDM table, but they chose not to. The environmental and carbon benefits had been achieved, and Shell was fine with that.

We presented the "Shell Sulphur Substitution Project" to a broad audience at the Globe Conference in Vancouver in March 2002, thus closing a very busy and colourful chapter and making room for the next.

Carbon and the Oceans

1999–2012

Virtually all current efforts to manage carbon emissions and climate mitigation are directed to the land base and, more specifically, to energy production, waste management, industrial and manufacturing processes, and forestry and agricultural operations. However, any meaningful look at carbon management or climate mitigation on our planet needs to take into account the role of oceans.

WHY DO OCEANS MATTER?

Oceans matter for a number of reasons. According to the IPCC, and as stated in its Fourth Assessment Report,[61] the vast majority of the planet's cycling carbon is found in the oceans. The amount of carbon residing in the oceans is approximately 45 times greater than that which is found in the atmosphere and 30 times greater than that found on land.

The approximate numbers are as follows:

- *Oceans*: 36,000 billion tonnes (largely in organic sediments)
- *Land*: 1,200 billion tonnes (most of it residing in living plant life)
- *Atmosphere*: 800 billion tonnes (most of it in the form of CO_2)

Collectively, this may appear to be a large amount of carbon, but it represents only a few tenths of one per cent of the carbon present at or near the Earth's surface.[62] Nonetheless, this cycling carbon is the carbon that we are seeking to understand and manage.

With respect to warming of the Earth, the IPCC in its Fifth Assessment Report[63] reports that "warming of the ocean accounts for about 93% of the increase in the Earth's energy inventory between 1971 and 2010 (high confidence)."

Given where most of the carbon resides, and where most of the Earth's reported warming is occurring, it is clear that oceans matter.

The IPCC's Fifth Assessment further noted that the rate of this warming is greatest near the ocean surface, where temperatures reportedly rose by approximately 0.11°C per decade between 1971 and 2010, or just under 0.5°C over the 40-year period.

CARBON ON THE MOVE

The world's oceans and the atmosphere are constantly exchanging enormous volumes of carbon at their interface in order to remain in relative equilibrium. These exchanges are accommodated by a complex range of physical, chemical and biological processes that dwarf anthropogenic emissions.

The IPCC reported in its Fourth Assessment Report that the annual release of carbon from the oceans to the atmosphere is estimated at approximately 88 billion tonnes. This annual release from oceans to the atmosphere is counterbalanced by an estimated 90 billion tonne transfer of carbon from the atmosphere to the oceans.

To put these natural emissions into perspective, the IPCC in its Fourth Assessment reported the annual CO_2 released into the atmosphere from the burning of fossil fuels to be 7.1 billion tonnes. Of this, 5.4 billion tonnes arose from fossil fuel combustion and cement production, and 1.7 billion tonnes from land use change, including deforestation.

In other words, anthropogenic carbon releases from the atmosphere were less than one-tenth of the natural releases from the ocean to the atmosphere.

WHAT DRIVES OCEAN UPTAKE OF CARBON?

The primary driver of CO_2 removals from the atmosphere is the photosynthesis of phytoplankton and other marine plants. In regions of the ocean where there is a healthy and abundant occurrence of phytoplankton, there will be a significant net flow of CO_2

from the atmosphere into the surface waters and then into the phytoplankton. If the phytoplankton are not consumed, they will eventually die, and the carbon-rich tissues will settle toward the ocean's floor, like leaves falling in a forest to join the soil below. Depending on the depth, some of the CO_2 bound in the plant tissues will go back into solution within the ocean water.

If the phytoplankton are consumed by fish or other organisms, then a portion of the tissues will be respired (burned), thereby releasing CO_2 back into the surrounding waters. Another portion will go into storage as new fish tissue and become part of the food chain.

Any fish not consumed by predators will eventually die and decompose, and the carbon-rich tissues will settle toward the ocean floor along with dead plant tissues. For example the White Cliffs of Dover are comprised of marine microorganisms that lived 150 million years ago.

In many regions of the ocean where there is an abundance of macronutrients but the occurrence of phytoplankton is diminished and productivity is low for some reason, there will be a net flow of CO_2 from the ocean into the atmosphere.[64]

These areas, largely devoid of life and representing up to one-third of the ocean's surface, are termed by scientists as "high nutrient, low chlorophyll," or "HNLC."[65]

EXPLORING OCEAN CARBON SEQUESTRATION

Given these carbon distributions, wherein over 90 per cent of the cycling carbon is found in the ocean domain, and given that HNLC regions represent a significant portion of the world's oceans, it should be no surprise that a few ocean scientists and forward-thinking entrepreneurs have been compelled to explore the possibility of unlocking the productivity of HNLC areas as a carbon management option.

This concept has been circulating since the 1930s but was finally given profile by oceanographer John Martin of Moss Landing Marine Laboratories in 1989, when he published the findings of some of his research in *Nature* magazine.

The key elements of Martin's theory were:

- Vast areas of open ocean have the macronutrients (e.g., phosphorous, nitrogen, silicon) required to be highly productive – but are virtually desert-like in terms of productivity.
- These HNLC areas, sometimes referred to as the "Desolate Zones," are largely devoid of life, having very little phytoplankton, which is the foundation of the food pyramid.
- The missing ingredient for HNLC ocean water is the micronutrient "iron."

Iron is naturally supplied to oceans by river runoff near the continents, the fallout from volcanic eruptions and deposits of iron-rich and transported aloft by storms from Asian deserts and deposited in the Pacific.[66] Many millions of tonnes of iron-rich dust are lifted annually from these deserts in storm events and carried aloft eastward over the Pacific Ocean.

EMULATING NATURE

Ultimately, the majority of this dust falls to the ocean, although significant amounts of Gobi dust can be found enriching the soils of some Pacific islands, such as Hawaii. Theoretically, where the ocean landing sites for the dust are HNLC, this natural supply of iron will unlock the suppressed productivity of the area, resulting in explosive algae blooms and dramatically amplified ocean productivity at all levels: from phytoplankton; to zooplankton; to small fish; to large fish; to pelagic birds; and finally to large ocean mammals.

In his theory, John Martin had simply illuminated the opportunity to mimic and augment a naturally occurring, random supply of terrestrial iron, delivered through storm events and subsequent iron depositions, with a controlled supply of iron, delivered by ship or aircraft.

When micronutrients are purposely delivered to plant life by the hand of man, in contrast to being delivered through somewhat random extreme weather events, the term used is "ocean fertilization," or in this case, "iron fertilization."

SOUTHERN OCEAN EXPEDITIONS – IRONEX I & II

In the early 1990s, John Martin planned the first iron fertilization mission, IronEx I, for testing in the "Desolate Zone" near the Galapagos Islands. Unfortunately, Martin passed away a few months before the research vessel *Columbus Iselin* set sail in 1993. The researchers introduced 450 kilograms of iron sulfate over two days over the targeted site, covering a 25-mile-square grid.

The experiment was a success, as it significantly and measurably increased phytoplankton production.

In 1995 IronEx II set sail. In Martin's absence, the project was led by Moss Landing's acting director, Kenneth Coale. On this second expedition the 450 kilograms of the iron compound were introduced over one evening.

Reports Charles Graeber from *Wired* magazine:[67]

> *And this time the dead seas sprang dramatically to life. Overnight, the* HNLC *waters clouded green. Fish were attracted by the harvest, and within days sharks and turtles were chasing the new food supply. By the end of two weeks, IronEx* II *had produced the biomass equivalent of 100 full-grown redwoods … a phytoplankton explosion of almost biblical proportions. The experimenters calculated that they had pulled 2,500 tons of* CO_2 *out of the atmosphere, and claimed they could do it again in desolate zones all over the world.*

Ultimately, dozens of universities and international research institutes became active in the space.

SOIREE AND SOFEX

The Southern Ocean Iron REIease Experiment (SOIREE) was an interdisciplinary study involving participants from six countries. SOIREE, which sailed from New Zealand in February 1999, was the first iron fertilization experiment performed in the polar waters of the southern ocean.

Approximately 5300 kilograms of iron sulfide were added to an area of 50 square kilometres. The resulting plankton bloom persisted for over 40 days following the researchers' departure from the site, as was clearly observed by NASA's SeaWiFS satellite orbiting 705 kilometres above.

Another southern ocean expedition (SOFeX)[68] in 2002, involving 76 scientists from 17 different institutions, sent three ships from New Zealand and Antarctica to study iron in the southern ocean. The study confirmed that the productivity of the southern ocean is largely controlled by iron availability, and that iron inputs have a much greater impact on carbon cycling than had been previously demonstrated.

John Martin's assertions[69] around the role of the natural airborne iron fertilization of oceans were largely confirmed in 2001. On April 10, Lawrence Berkeley National Laboratory launched, from the US Coast Guard's icebreaker *Polar Star*, two deep diving robotic "carbon observers" at Station PAPA, 1,000 miles (1600 kilometres) west of Vancouver Island, British Columbia.[70]

The robotic "carbon observers" were measuring a range of biophysical and chemical parameters, as they transited daily from the ocean's surface to a depth of 1000 metres and reported their findings to satellites orbiting above.

The carbon observers were launched three days after a large storm on the Gobi desert lifted iron-rich dust into the jet stream headed toward the Pacific. On April 12, some of the dust fell from the sky and deposited on the waters of Station PAPA. The carbon observers reported a doubling of particulate carbon in the water column. These findings were corroborated by NASA's SeaWiFS satellite, which peered through clouds to see an ocean surface turning green with plankton.[71] These measurements provided the first direct observation of windblown terrestrial dust stimulating ocean productivity.[72]

OCEAN CARBON FROM SPACE

Two years earlier, a few colleagues and I had established Ocean Carbon Science Inc. (OCS), as a specially purposed collaborative initiative that would take on a few specific carbon measurement challenges. OCS began to take shape during a timely dialogue with John MacDonald in a parking lot at the University of British Columbia. John is the Order of Canada scientist and engineer who, as mentioned in Chapter 4, co-founded MacDonald, Dettwiler and Associates (MDA), a global leader in aerospace technology, including the designing, building and operating of satellite systems and land stations.

Earlier that day, John and I had participated in a round table discussion at UBC's Sustainable Development Research Institute (SDRI). Recalling our conversation, as we stood in the parking lot, I asked John if he was interested in having satellite technology applied to measuring ocean health,[73] and more specifically with respect to measuring the carbon dynamics of oceans. He confirmed he was always trying to find more applications for his "birds," and in retrospect that is in part why he was attending the SDRI meeting, which had carbon and climate on the agenda.

In the parking lot he found what he was looking for, and I found a like-minded collaborator.

An immediate challenge for OCS was to develop methodologies to accurately quantify ocean-based CO_2 uptake in faraway open ocean regions. My graduate research on carbon sequestration in trees a decade before began calling to me with a potential piece of the answer. I had mused then about the possibility of equipping satellites with appropriately tuned fluorescence sensors, in essence modified versions of what I had been using, that could "see" photosynthetic carbon uptake in trees and forests. But I had parked the notion on graduating.

John Woodruffe with whom I had worked in China, was another participant in OCS. John was closely linked to the fossil energy industry, which might have an appetite to support such work – particularly if a supply of credible, multiple-benefit[74] carbon offsets might be a future outcome.

For a period, LENR (Low Energy Nuclear Reactions) investigator Russ George was involved in OCS but went on to found a research vehicle of his own, Planktos.

The OCS principals understood that fungible carbon offsets arising from oceans were likely a long way off, if possible at all, but we all agreed that the ocean piece of "the carbon equation" was too large to ignore. We needed at least a few strong partners to finance the work. This was a challenge. When we approached Shell Canada, the response was clear. The international Shell parent organization in Europe was still licking its wounds following the Brent Spar incident in 1995, when a large NGO reportedly decided to target Shell over an oil rig that was being decommissioned. Shell was preparing to carry out the recommendations of the consultants, but before they could do so an environmental protestor chained himself to the rig. A young female pilot then flew a chartered helicopter perilously close to the rig, presumably to create photo opportunities. The images would soon be seen on front pages around the world[75] a few days later, and Shell was immediately painted with a black brush for which there was no thinner, except time.

Clearly Shell was "out" as a potential sponsor, but we were able to secure the interest and support of BC Hydro, British Columbia's Crown power utility, and Edmonton Power, a city-owned power utility using primarily coal to generate electricity. Both had strong sustainability policies. Within that context and their broader interests in supporting carbon management research, they agreed to support a research and measurement-focused project. The work would in part be undertaken by those universities having the scientists and advanced capacity required to study carbon, energy and micronutrient flows and dynamics in complex ocean systems. OCS would support and coordinate aspects of the research pertaining to remote sensing and measurement and possibly, at some point, carbon offsets. In taking on this role, OCS became the lead industry partner in a broad collaboration, supported by the Natural Sciences and Engineering Research Council of Canada and several universities.[76]

OCS also began establishing relationships with other organizations such as the Electric Power Research Institute (EPRI) in Palo Alto, California, and an aquaculture enterprise out of Tahiti, reportedly having the support of the country's president, Gaston Flosse, with a business model incorporating the sequestration of ocean carbon into the commercial production of cultured pearls.

TROUBLES IN WASHINGTON

Unfortunately, while we were busily establishing our relationships and were in the process of defining programs of research, trouble was brewing.

It turned out that there was significant pushback against doing any industry-backed research from some green organizations and a number of academics, including an outspoken MIT professor. This became abundantly obvious at an ocean fertilization science and policy workshop I attended that was organized by the Association for the Sciences of Limnology and Oceanography (ASLO) in Washington, DC, in April 2001. The distrust of industry participation was palpable.

This caution was representative of a broader sentiment against what was sometimes referred to as "geo-engineering," even though that term was generally associated with grand-scale ocean manipulation, as opposed to the small-scale, mimic-nature, measurement-focused research we were preparing to undertake.

Frustrated and looking for help, I reached out to Carlton Bartels, who was director of emissions trading for Cantor Fitzgerald, headquartered in New York City. Carlton was a recognized visionary in the nascent carbon trading space and had been discussing the possibility of "carbon credits" as early as 1980. I hoped he might provide some useful perspectives on the challenge we faced.

In early summer of 2001 Carlton was travelling through western Canada to participate in a series of carbon emissions trading exercises with the industry and government players in that region, primarily the energy majors in Alberta. He would be visiting Vancouver.

The OCS offices were located in the World Trade Centre in Vancouver, and Carlton was staying at the Pan Pacific Hotel next door. We met there over breakfast. Carlton quickly understood the situation. He had no immediate answers, but we agreed to give it some "soak time" and reconnect in the fall.

A few months later, on the morning of September 11, 2001, I was discussing the growing anti-industry sentiment of researchers with Jonathan Phinney, executive director of ASLO)[77] in Washington. It became apparent that given the structure and resourcing of the OCS initiative, it might not be able to complete its work – certainly not on a schedule that would keep its industry partners, who were understandably sensitive to the optics of being associated with potentially controversial research, at the table.[78]

Ours was a surreal discussion on a day no one alive at that time will forget. As we chatted, Jonathan described a curious swirl of dark smoke he saw rising upward from the Pentagon. I had no way of knowing that Carlton Bartels and much of his team, who were working at his Cantor Fitzgerald office on the 101st floor of the World Trade Centre that same morning, had already perished with another 3,000 souls as a passenger aircraft hijacked by terrorists flew directly into their tower and exploded.[79] The Pentagon had been another target.

Months later, once recovered from the shock of 9/11, we would continue to seek support for the OCS initiative. But in the absence of "patient capital," the discontinued support of the energy companies, and with no hope of support or guidance from the decimated industry giant Cantor Fitzgerald, we chose to wrap up OCS and get on with our respective lives.

This was a great disappointment. In particular, the opportunity to work with Professor Dave Karl, leader of the SOEST Laboratory for Analytical Biogeochemistry at the University of Hawaii at Mānoa, was slipping away. Dave, with whom I had met on several occasions in Hawaii, as well as at conferences in Washington, DC, and Vancouver, was clearly a world leader in investigating ocean chemistry and the HNLC phenomenon. He had been very generous with his time and was fully prepared and equipped to co-operate and undertake research that would support our proposed investigations. The message I prepared on behalf of OCS and its partners to inform this accomplished SOEST leader of our predicament, and the stark consequences (no support for ocean research), was one of the most difficult letters I have ever written.

In any case, that was over 15 years ago, and it is clear that the generation of validated, verified and registered carbon offsets from ocean-based sequestration, if ever possible, will require significantly more time, effort, resourcing and education. Nevertheless, some other groups and individuals have continued their work, approaching it from a somewhat different angle – fish production.

THE CAT CAME BACK

Like the proverbial cat that came back, Russ George simply wouldn't quit. He didn't know how, and continued undaunted. Russ, operating from his not-for-profit organization, Planktos, and enjoying some notoriety from a story on his work in the journal *Nature*,[80] just kept going. In 2005, Russ secured the use of *Ragland*, a 101-foot Baltic trading schooner built in Denmark in 1913 to haul granite. *Ragland* was owned by the Canadian songwriter and rock star Neil Young,[81] who happened to also own a ranch in the area and had his boat moored next to Russ's at a marina an hour or two south of San Francisco.

In June 2005, Russ loaded *Ragland* with several bags of red ochre pigment, set the sails and headed west toward Hawaii. A week or so later, in international waters, he introduced the iron-rich pigment to the ocean. Russ claimed that this fertilization "removed the CO_2 content from the atmosphere of about 3,000 households."[82]

Two years later, Russ accepted an invitation from a Canary Islands university to conduct joint research activities involving ocean fertilization. Russ secured the 110-foot research vessel *Weatherbird II* and set sail east from the US in November 2007. However, when the ship reached Spain, the government refused to let it dock, and the expedition sputtered.

ONE MORE TIME

In 2012, Russ undertook yet another run at ocean fertilization, this one off Haida Gwaii, approximately 800 kilometres north of Vancouver. There he had partnered with First Nations, acting as the chief scientist to the Haida Salmon Restoration Corporation (HSRC). His stated intention was to restore salmon returns to the First Nations communities, from an ocean region of somewhat diminished productivity. Russ was reported to have presold carbon offsets to an airline tycoon in Germany. For this project, Russ had secured yet another boat, the commercial fishing vessel

Ocean Pearl. Russ began to prepare for his biggest ocean fertilization effort ever.

STORMY WEATHER AND THE WORLD'S FIRST CARBON RAID

Days of stormy weather notwithstanding, in the summer of 2012 Russ and his crew headed north and proceeded with the release of 180 tonnes of iron sulfide over a 10,000-square-kilometre area situated west of Haida Gwaii. This time the team was able to deliver. The plankton bloom was clearly visible from space by satellites and was reported to have attracted "herring, salmon, tuna, dolphins and even whales ... to feed over the next couple of months."[83]

Months later, when officials from the federal government of Canada learned details of his actions, which they considered might have been in contravention of an international convention and/or federal laws, they took action.

Reports Russ:

> *Just days before Easter our small village research office in Vancouver was swarmed by 11 officers dressed in all black combat gear, armed and imposing beyond belief. In the largest assault/raid in Environment Canada's history the* RAID *team rushed into the office of 7 people, counting the 2 grandfathers and 2 women present. They barked forceful concise commands and shouted out asking whether there were any weapons or dangerous tools in the office.... Herding the 7 people in our office down to our basement they boldly asserted that we had to comply with their every demand or*

face unpleasant consequences.... They finally left 24 hours later at around 8 a.m. the next day and allowed us to resume business, such as it was.

Reportedly, Environment Canada had warned the project organizers that they would need to apply for a disposal-at-sea permit. After the company finished its mission at sea, without said permit, the ministry launched an investigation into potentially illegal ocean disposal. The organizers fought to have a search warrant declared invalid, but to no avail. To date, no charges have been reported.

Meanwhile, back in the waters west of Haida Gwaii, it appeared something tangible, and something good, might have occurred. NASA satellite images clearly showed a large plankton bloom in the experimental area, as would be expected in the case of a successful project.

Furthermore, salmon returns were dramatically "up" from previous years. Reports *Planet Experts*:[84]

The largest run of Pink salmon occurred between 12 and 20 months after the ... iron seeding ... the 2013 pink salmon harvest was the second most valuable on record ... salmon catches have surged from 50 million to 226 million.

Anyone familiar with west coast salmon fisheries is aware that dramatic changes in returns, up to an order of magnitude between years, are not entirely uncommon.[85] And so, while it cannot be concluded from these salmon harvest numbers that the fertilization was successful from a fish enhancement point of view, it is certainly within the realm of possibility. Currently, the government of Chile is investigating the use of iron fertilization to recover its fisheries, apparently without thought to carbon.

Given the possibility of fisheries enhancement success reported above, the sheer magnitude of carbon cycling and storage in the oceans, the implications of same with respect to carbon management and climate policies and strategies, and an array of environmental sensitivities, it would appear that the "Carbon and the Oceans" story is very far from being over.

The Millennium Odyssey

2001

It was the second springtime of the new millennium, and I was chatting over tea with a wise friend. I had met Michaela over 15 years before, lost track of her and then reconnected in 1998. She seemed to appear in my life as a wise soul when there were significant issues to consider, and had been integral to my decision to undertake my PhD back in 1986. "This is a wall you must walk through," she had suggested – strong wording from someone who rarely gave advice, but rather provided rich, deep and often humorous perspectives.

I had taken Michaela's counsel and walked through the wall. With that decision long behind me, I was explaining to her a predicament I had found myself in that was very uncomfortable.

CLEAN ENERGY: LOW ENERGY NUCLEAR REACTIONS (LENR)

Several years earlier, back in the Westcoast Energy days, I had taken an unusual assignment from a chief officer of the company. This began with a *New York Times* news clipping he had distributed to the company's senior management. The story was about revived interest in Low Energy Nuclear Reactions (LENR). The story followed from extraordinary reports of test tube-scale, zero-emissions nuclear reactions discovered at the hands of scientists Pons and Fleischman which had been reported globally as "cold fusion." After the initial clamour, the reports were dismissed as bogus, and Drs. Pons and Fleischman exited their US laboratory to live and continue their work in France, supported by Toyota Corporation.

The *Times* story presented evidence that work on LENR was continuing and that there might be something there after all. While most mainstream scientists dismissed the story, given the potentially very low cost of energy generation with zero radiation, a few people in the established hydrocarbon energy industry saw LENR as a potential threat to the existing power generation infrastructure. My chief officer wanted to know if it was something the natural gas industry should be concerned about. I agreed to go on a mission, quietly, to delve into the matter.

In the spring of 1995 I flew to Palo Alto on a fact-finding mission. I quickly learned that some credible organizations, such as the multi-billion-dollar Electric Power Research Institute (EPRI) located in Palo Alto, and Stanford Research Institute (SRI) International located in Menlo Park, were involved.[86] However, I came home unconvinced, and I reported back that there was "nothing imminent."

However, while nothing appeared imminent, I also knew that *if* this apparent breakthrough was real, as in being repeatable, controllable and scalable to commercial applications, it needed to be understood and prepared for. If LENR fusion could produce ridiculously huge amounts of energy with a small tumbler of heavy water, at a very low cost and with no radioactive waste, then the world's geopolitics would change. Dramatically. Perhaps dangerously.

A CHANCE ENCOUNTER

It was a very large "if" but one that I thought ought to be explored. An opportunity arose as I explained the situation to Michaela. Michaela was renting a room in a home in Deep Cove, North Vancouver, owned by a gentleman named Spar Street. As I recounted and as she listened, I heard some rustling in the kitchen, and then a voice. "Maybe I can help."

In walked Spar.[87] Tall, blond and articulate, Spar was a very successful commercial artist. His paintings were displayed on the walls of the United Nations building in New York. And as it turned out, Richard Branson also valued Spar's art, some of which hung in Branson's home back in Britain. When in Vancouver, Branson would sometimes have dinner with Spar at his home in Deep Cove, at the table where we were now enjoying our tea.

Richard Branson was celebrated as a man of vision, living out of the box – and Spar was quite certain that Branson would be interested in this development – again, if it was real. Spar was prepared to make the connection.

This was getting interesting. I had recently reconnected with Brian Anderson, who had been the chairman of Shell North East Asia when I was first working in China in the late 1990s and early 2000s. Shell was well known for its scenarios planning work, which looked forward many decades at various potential energy scenarios and their impacts on geopolitics. Brian listened to my LENR story and suggested that even if it had only a slight chance of being real, given the geopolitical implications, a scenarios exercise was warranted. Brian knew Shell's lead scenarios guru and would connect us with him if and when we wished him to do so.

About the same time, I had connected with Sir Arthur C. Clarke. Arthur was an extraordinarily talented and prolific British science fiction writer, futurist, inventor, undersea explorer and television series host. He foresaw the computer revolution decades before it arrived, as featured in the introduction of the film *Steve Jobs* released in 2015. His home was now in Sri Lanka. I had first learned of Arthur C. Clarke from watching the film *2001: A Space Odyssey*, which he co-wrote. This is considered to be one of the most influential films of all time, and it was but one of a plethora of Arthur's accomplishments in writing and film.

And here I was in 2001, the first year of both the 21st century and the third millennium, communicating directly with him by email and snail mail.

Arthur was another visionary and an early promoter of the geostationary satellites that would eventually accommodate mankind's needs for international telecommunications and global positioning systems. Arthur began designing these systems in the 1940s. In his mind's eye, these satellites would be powered by nuclear energy generated by small on-board reactors.

Within that context, Sir Arthur was very interested in LENR. And Sri Lanka, we agreed, looked like a suitable location for a scenarios workshop. Sir Arthur was open to hosting a visit and posted me a photo showing the natural beauty of the area, with a handwritten note attached saying, "See what I mean?" And so, 23 years after abandoning my backpacking plans, it appeared I would make it to Sri Lanka after all.

ART IN THE BIG APPLE

Things were lining up for something special. We had a potential venue in Sri Lanka. I knew some of the researchers in LENR and had the interest of Brian Anderson, who in turn had the Shell scenarios planner within his circle. Spar seemed rather sure that Branson, a respected entrepreneur, would be interested in such a workshop given the potential climate mitigation potentials of a zero-emissions technology. And of course there would be business opportunities.[88] It turned out Spar was soon to be headed to an art show in New York City, where he expected to meet a lieutenant of Branson's. And, in another amazing piece of serendipity, Brian Anderson was headed to New York City to serve as Executive in Residence at Princeton University during the same week Spar was going to be there for the art show.

I connected with Sir Arthur by email, and sure enough, he

confirmed the invitation to his Sri Lankan home for a workshop. When would we like to come?

I sent Branson a letter and booked a March 2001 flight to New York, where I would stay with Spar. I would connect with Brian over dinner and start to lay the plans, and Spar would confer with Branson's lieutenant. Then we would all get together and hatch a plan and schedule for the Sri Lanka workshop. Perfect.

It was a crazy time in New York. The city was hosting a large international art show that covered acres. Spar would be displaying there. Each night, absurdly stretched, quasi-assault vehicles, more specifically "Hummer" limousines, would cruise around the city, carrying artists, promoters and of course the supermodels. This went on throughout the night. It was a dazzling experience.

We were in for a few surprises, however.

Brian Anderson, whom I had last seen in Beijing when working on the Shell project, was to meet me at an Indian restaurant in the Soho district. Brian arrived a little late, apologized and went on to explain what had happened. A polished and highly informed lecturer, Brian had finished his visit and was attending a reception at the university for the future mayoral candidate Michael Bloomberg, when a woman approached him with some documents. It turned out Brian was being served a summons with respect to Shell's alleged involvement in the execution of Ken Saro-Wiwa and eight other Ogonis in Nigeria.

Saro-Wiwa was a Nigerian writer, television producer and environmental activist. It turned out that Brian had been acting as chairman of Shell in Nigeria in 1995 when the Nigerian government executed Saro-Wiwa for allegedly masterminding the murder of a number of Ogoni chiefs at a pro-government meeting.

I was stunned. But Brian was calm as he explained the situation to me. Having worked with the man, who in my experience was of the highest integrity, and knowing virtually nothing myself about what happened in Nigeria, I simply listened.

Then we turned the discussion back to the proposed workshop.

FUTURUS INTERRUPTUS

The next surprise was that Spar had not been able to make contact with Branson's New York lieutenant. There was no explanation, but I surmised that Branson had required that some due diligence be done on the proposed workshop in Sri Lanka. As "cold fusion" was considered a dodgy area at that time, he understandably needed to be very careful about any perceived association.

Without Branson, the impetus to proceed with the workshop waned. We believed we needed the entrepreneurial component, and had apparently lost it. He was not easily replaced. At the same time, I was getting deeply involved in a wind energy project[89] in the southwestern United States, working on the avoided carbon values, and this was taking most of my time.

The party was over, at least for then. Sir Arthur C. Clarke succumbed to respiratory failure in 2008 at the age of 90,[90] and I had lost touch with the LENR scientists. Brian Anderson was (and still is) shuttling back and forth between France and Hong Kong, and Spar Street, my connection to Richard Branson, had taken up residence in Maui.

It is unlikely that Branson could have anticipated that 15 years after he declined our workshop, Carl Page, brother of Google co-founder Larry Page, would be supporting further LENR research at SRI International in California, through his company Brillouin Energy. Another LENR lab had been established in 2015 in Japan by Vancouver-based philanthropist Hideki Yoshino. Located at Tohoku University, Yoshino and his partners secured the participation of the Japanese government as well as Mitsubishi Heavy Industries. Also getting involved were Nissan and Toyota.

Arguably, from the vantage point of the first quarter of 2017, the LENR story has really just begun. And it would appear that the need for a high level scenarios planning exercise that addresses as well the accelerating swing toward a low carbon economy is now greater than ever.

"Build It and They Will Come ... Eventually"

2003–2011

Following a two-year exploration of the potential for carbon sequestration in oceans and involvement as an investor and consultant in several "clean energy" enterprises, I had an encounter that marked the beginning of a journey back to trees and photosynthesis. And yet another return to carbon.

During the Christmas season of 2003, I met socially a gentleman named Bart Simmons. Bart had earned a master's degree in nuclear engineering from the University of Toronto. Bart was very bright and continued to consume literature in the field of Low Energy Nuclear Reactions (LENR), but his true passion was forest restoration. In fact he had operated a number of reforestation companies that had planted hundreds of millions of trees across Canada, over several decades.

Bart was smart. He was charismatic. And he was highly entrepreneurial.

When I met Bart he had recently become interested in the business potential of the carbon offset trading world. Like the *New York Times* reporter back in 1989, he had joined the dots between trees, carbon and possible climate mitigation. And he saw a business opportunity. After a false start in the forest carbon space with a few other players, he knew that to succeed he needed to partner with genuine experience and capacity in climate change and carbon markets to complement his reforestation know-how.

THE THWARTED 2010 OLYMPIC LEGACY FORESTS

Bart and myself hit it off like a house on fire, and after a few bumps, ERA Ecosystem Restoration Associates Inc. was born. We became operational in June 2004. With Cornelia Rindt, a very enthusiastic and capable intern, to help us, we went to work developing project ideas in Bart's basement in West Vancouver. Early in the company's development, we decided that we would go after the prize of providing the carbon offsets for the 2010 Winter Olympic Games, to be held in Vancouver and Whistler. Based on this dream, and the belief that other folks would sign up for ERA's forest-based carbon offsets, we raised some private capital and set the wheels in motion.

The founding business model was simple, or so it seemed at the time:

- secure 100-year agreements with local municipalities that would have ERA restore their degraded forest ecosystems in return for acquiring the rights to the "carbon offsets" that would be generated from restored ecosystems as they gobbled up CO_2[91]

- develop prescriptions and replant native tree species in degraded urban and rural forests, with the added benefits of restored native biodiversity and habitat, at no cost to the municipality
- validate, verify and ultimately register the carbon offsets
- sell the "carbon offsets" to the 2010 Olympic Games and other buyers that wanted to be carbon neutral, climate friendly, etc.

Armed with this vision, and supported by the release of the United Nations Millennium Ecosystem Assessment,[92] we met with and presented our pitch to a number of mayors and councils in the Lower Mainland, and ultimately were able to establish several 100-year agreements with various municipalities in the Metro Vancouver area, the first being Maple Ridge. And on Rivers Day, September 29, 2005, with the help of the mayor of Maple Ridge and some of ERA's clients, friends and family, we kicked off with our first plantings of native trees in a Maple Ridge park area adjacent to the Alouette River.

Of course there had to be a glitch, and it didn't take long to show up. It came in the form of a "cease and desist" type of letter from the 2010 Olympic Games legal team. The games lawyers took issue with our using the word "Olympic" as in "Olympic Legacy Forest," in any of ERA's communications. Apparently they held some sort of international rights to the word, and we were not authorized to use it.

Flummoxed but not deterred, we then offered to offset the operational CO_2 footprints of a handful of forward-thinking organizations, both big and small. ERA's first clients included an auto lease company (AutoOne), Sustainable Produce Urban Delivery (SPUD), the Globe Foundation of Canada and BC Hydro.[93] Eventually Shell Canada and others would join the party.

Creating and selling carbon offsets was a difficult business. The very notion of "carbon offsets" was poorly understood and still is. Beyond that, the work was labour intensive and expensive. Restoration prescriptions required the services of a professional ecologist, while complex carbon calculations and computer model building required qualified scientists, and implementation required skilled forest workers. Validation and verification services were difficult to secure, expensive when we could, and the "standards" that would ensure quality to buyers didn't yet exist.

There wasn't a lineup of buyers. To many, carbon offsets looked like snake oil.

"GOING PUBLIC AT THE BRINK..."

We persevered. Up to this point we had financed the company through a few private investors, friends of family, and Bart and I dipping into our back pockets and lines of credit to deal with cash calls. In 2007, my friend Warren Carr, a seasoned production manager from the film industry and a huge advocate of the EcoNeutral© brand ERA had built, became involved. Through his contacts, the film, sports and entertainment industries entered the picture. Through Warren's foundational work, ERA was able to become a provider of carbon offsets for a number of musicians, including Neil Young, Sheryl Crow and Sarah McLachlan. ERA also supplied carbon offsets to a number of politicians, including Al Gore, Bill Clinton and Canadian Senator Elaine McCoy. And through the not-for-profit Zerofootprint organization, ERA's offsets were used by Air Canada's voluntary offset programming for travellers.

Warren introduced me to Graham Harris, a public markets veteran who had taken a number of companies through the "going public" process. Graham immediately saw the potential for ERA in the public markets world. Ultimately, he and his colleagues in Toronto convinced us that as a public company, ERA would be able to raise the capital required to scale up our production for anticipated markets, and therefore we ought to do it. This made sense.

Taking the public vehicle route required many trips to Toronto. Out came my dusty suits (Bart didn't "do" suits), and off we went. In Toronto, a modest limo would pick us up at Pearson Airport, and soon we were trotting around the financial districts with our dog and pony show. Big and small, we saw them all.

A typical day would have our Toronto agent in his $2,000 suit, me in my $200 suit and Bart in his jeans, snappy sports jacket, and snakeskin boots pulling his cart with who-knows-what stored in it,[94] as we made ERA's presentations to wealth managers and hedge funds.

The meetings, most often conducted in an empty boardroom, sometimes adorned with highly artistic photos of oil rigs and open pit mines, would begin with introducing me (CEO), Bart (COO) and the opportunity. Then I would walk through the PowerPoint presentation, while Bart invariably worked on his Blackberry, under the table. When it was time to add some technical eco-speak to the discussion, I would nudge or kick Bart under the table, and he would spark up and speak eloquently to the minutiae of ecological restoration, until the finance guy's face glazed over. When it came to ecological restoration talk, Bart was flawless.

Over several months in the spring of 2008, we made countless pitches. Thankfully, some folks stepped up to the plate.[95]

On July 26, 2008, ERA Ecosystem Restoration Associates Inc. completed an RTO (reverse takeover) with Interim Capital, a public vehicle, and became ERA Carbon Offsets Ltd., listed on the TSX Venture Exchange as "ESR."[96] Valued at $17-million, ERA was the first publicly trading carbon offset company in Canada. Trading on the TSX Venture Exchange began on July 26, 2008.

Within days of this launch, the financial world was turned upside-down by a crisis that continues to be felt around the globe to this day. Fortunately, the ERA vehicle was built, it was resourced, it had a growing inventory of carbon offset product to sell, and the company was ready to rock.

CARBON "POLICY RISK" IS MANIFEST

But there was a problem. Notwithstanding the fact that we had built it, and they were coming, ERA's inventory continued to far exceed the demand. More importantly, a clear path to profitability was not apparent. Perhaps most significantly, the pending US carbon cap and trade program (defined by the Climate Stewardship Acts) which figured largely in ERA's business plans and for which we had lobbied in Washington, was being derailed in the House of Representatives by the Republicans.

A multi-billion-dollar market, one that was integral to ERA's business plan, had just gone "poof," and with it went any certainty regarding the future of national-scale carbon offset markets in North America.[97] This uncertainty was reflected in the Chicago Climate Exchange, launched in 2003 by financial futures guru Richard Sandor, ending its carbon trading activity in 2010.

While there were other initiatives afoot that would make markets on a regional scale, such as the Regional Greenhouse Gas Initiative (RGGI) in the northeast states, the Pacific Carbon Trust (PCT) in British Columbia and the Western Climate Initiative (WCI) in some western states and Canadian provinces, the rules did not accommodate the forest ecosystem restoration based carbon offsets that ERA was producing. And while Alberta's Specified Gas Emitters Regulation could conceivably accept CERP-type carbon offsets, the regulation excluded carbon offsets supplied from outside the province.

SAVED BY THE BOOTS

Fortunately, thousands of miles north and west, in July 2009, near Bombay Peggy's Brothel in Dawson City, Yukon, a particularly fine pair of white cowboy boots and a carbon cupid were about to come to the rescue.

Weiße Cowboy-Stiefel (White Cowboy Boots) in the Yukon

2009

The summer of 2009 saw ERA in an admirable but challenging position. The company had become publicly trading a year before, in July 2008, just weeks before the collapse of stock markets in Canada and around the world. We had successfully raised $4.5-million for operating capital and to prepare for the anticipated US cap and trade system.

As well, we had in our inventory over 200,000 tonnes of validated and verified carbon offsets arising from ERA's Community Ecosystem Restoration Program (CERP) and were generating more monthly. We also had some prestigious clients and users of our product, including some high-profile politicians and rock stars.

However, the numbers did not add up. Given the then current

pricing of offsets, the sale of our offsets fell far short of addressing our operating costs, and we were drawing down the treasury quickly. On the horizon, we had believed, was a regulated US cap and trade system that could truly launch the company. However ERA's Canadian-based voluntary carbon offsets would not be fungible in the anticipated regulated US trading system, and our offset sales volumes barely put a dent in ERA's inventory of product.

MEANWHILE, UP IN THE YUKON

Then, in July 2009, the forest carbon world changed. It began with a pair of white cowboy boots navigating the muck and mire into a historic brothel in Yukon's Dawson City.

Erin Kendall was a bright, entrepreneurial young lady with solid experience in forest restoration, whom we hired to help manage new CERP projects. She was somewhat of a free spirit, and sometimes, if not most of the time, I didn't have a clue where she was. In July 2009, she was not to be seen – unless, as it turns out, one was in Canada's Yukon.

There Erin could indeed be found, attending the annual music festival held in Dawson City. On the second evening of the festival, which was besieged by a cold rain, Erin had settled in for a warming beverage at "Peggy's," a former "place of ill repute" no doubt blessed with a thousand colourful stories – and one more gestating. As Erin recalls:

> *I didn't really know why I was there again – my second year in a row at this particular festival, 40 hours overland one-way from Vancouver. I just knew that it had felt*

important to get back there no matter what, and I had talked a few other fools into joining me.

After Peggy's, Erin moseyed on up to the beer garden bar down the street a ways. Standing in line for a beverage, awash in mud, Erin felt a presence behind her.

I turned around to see who the offending party might be, and spun quickly back to face forward, heart caught in my throat.... As we inched forward through the mud he said he liked my boots, the only white cowboy boots to be seen in a sea of sensible hikers.

Erin and Dennis de la Haye began to chat. And dance...

MADE IN CANADA

Dennis was a brand strategist visiting from Germany with his father, five days fresh from Berlin. Dennis worked with some large firms, including Lufthansa, Volkswagen and others, that might just be looking for some "Made in Canada" branded forest carbon offset products. Erin and Dennis saw the potential to do some business, returned to their respective corners of the planet and began strategizing. A larger dance had just begun.

Dennis identified a German energy company, HEAG Südhessische Energie AG (HSE) Entega,[98] that was looking to develop and market a carbon-neutral natural gas product. Apparently the notion of offsetting their gas products' greenhouse gas footprint with made-in-Canada forest carbon offsets was appealing to them. This was confirmed when a number of well-spoken German folks showed up at ERA's North Vancouver offices, to "kick the tires" of our operations up the Fraser Valley. We showed the German team our projects in Mission and Maple Ridge, and introduced them to our key partners, including the mayor of the City of Langley, Peter Fassbender, who was a strong supporter of our projects.

The German team went home pleased, and soon thereafter we were negotiating large-volume sales of ERA's carbon offset products at a price we believed was workable.[99] But the plans did not stop there. A director at HSE Entega had a larger vision.[100] Following a third-party due diligence undertaking with respect to ERA, he and his associates decided to form a new company. The "Forest Carbon Group" would be established to secure a supply chain of qualifying carbon offsets, ostensibly from ERA and possibly other suppliers, and resell them to Entega and thereafter other large clients in Germany. As well, the Forest Carbon Group would take a 29 per cent equity position in ERA. This investment would put more capital into the ERA treasury. And it had other implications we hadn't thought of.

The Germans were clearly enchanted with the Canadian-sourced ERA carbon offset product. At the annual HSE Entega 2009 Christmas party held in two warehouses in Mainz, Germany, the entire motif was Canadian, with a full-on dogsled team parked at the entrance, mounted moose antlers hanging over mice-ridden sofas,[101] with bannock and cold fish soup being served as appetizers.

As ERA CEO, I was called to the stage and introduced by Managing Director Holger Mayer to the large gathering of employees. It was a unique experience seeing a German version of Canadiana.

As a consequence of HSE Entega's appetite for carbon product, ERA was quickly moving from inventory-rich to inventory-light, and we were under pressure to produce more product, faster. Over the following years, on behalf of Entega, the Forest Carbon Group bought almost every tonne of offset product ERA was able to have validated, verified and registered on the Markit Carbon Registry.[102]

ERA responded by expanding its reach. The company undertook an ecosystem protection project in BC's Gulf Islands with the Ministry of Environment,[103] and began developing the Mai Ndombe project with forest communities in the Democratic Republic of Congo. During the period 2009–2011 ERA's sales represented close to 10 per cent of the entire global voluntary forest carbon offset market.

The white cowboy boots in Dawson City had triggered a course of events that had a significant impact on the global voluntary forest carbon market – surely something that could never have been anticipated when Erin slipped on her boots and sashayed up to the bars of Dawson City.

Lunch with Al Gore and "The Colossal Blunder"

2008–2015

One of the most significant sources of CO_2 emissions found in the atmosphere is deforestation. With tens of millions of acres of rainforest being destroyed annually, it has been estimated that deforestation now accounts for about 10 per cent of annual carbon emissions – an amount approximately equal to the total emissions from Western Europe.[104] At the same time, the toll that deforestation has and is taking on wildlife populations, biodiversity and other valued ecosystem services is staggering.

GREEN MISSION: LOST?

Arguably, a significant shortfall of the Kyoto Protocol established at the 3rd Conference of the Parties (COP 3), and as operationalized through the Marrakesh Accords in 2001 (COP 7), was the failure to provide an adequate foundation for reducing carbon emissions by arresting deforestation in the developing world. This situation arose in part from the success of some green advocacy organizations in largely excluding the protection of rainforest ecosystems in developing countries from international climate change policy. This remarkable irony seemed to have been missed by these organizations, whose members appeared to have temporarily lost sight of where the word "green" came from – the colour of the foliage of living plant life and standing forests. For many if not most of the non-governmental environmental organizations, "green" had become a metaphor, its original meaning lost. Photovoltaic panels, wind turbines and electric cars were considered "green" and worthy to promote and include in climate mitigation strategies – while projects that protected rainforest ecosystems apparently were not.

A "COLOSSAL BLUNDER" RECOGNIZED

Recognizing this significant failure in Kyoto, *The New York Times* published on May 29, 2009 a strongly worded editorial titled "Forests and the Planet":

> *A major shortcoming of the 1997 Kyoto Protocol on climate change was its failure to address the huge amounts of greenhouse gas emissions caused by the destruction of the world's rainforests....*
>
> *The planet has been paying for that colossal blunder ever since.*

This editorial echoed an awakening and call to action eight months earlier by some of the organizations that had been part of the Kyoto experience leading to the "shortcoming." There had been a realization that their original objectives vis-à-vis native biodiversity and fish and wildlife habitat protection were linked to *protecting* forests, not pushing them out of the policy equation. This fundamental shift in perspective was manifest on September 22, 2008 at a luncheon held in New York City, where a number of senior government ministers, key international environmental organizations, the banking community and carbon players were invited to come together to illuminate and address the problem.

Led by Jeff Horowitz, the high-energy leader of Avoided Deforestation Partners, and moderated by former CBS news anchorman Dan Rather, the meeting at the Yale Club venue on Vanderbilt Avenue featured virtually all of the big players in the carbon world. Former US Vice President Al Gore and (the late) Dr. Wangari Maathai, both Nobel Laureates, spoke passionately about the potential roles of forests in climate mitigation, and their assertions were corroborated by declarations of support by the senior officers of key international organizations, including: Carter Roberts, CEO of the World Wildlife Organization (US); Dean Hirsch, President, World Vision International; Helene Gayle, President, CARE USA; Raymond Offenheiser, President, Oxfam America; Kevin Knobloch, President, Union of Concerned Scientists; Kevin Conrad, Director, Coalition for Rainforest Nations; Mark Tercek, President and CEO, The Nature Conservancy; and Peter Seligmann, Chairman, Conservation International.

Major financial institutions such as Merrill Lynch, Macquarie

Bank Ltd., the Bank of America and Goldman Sachs participated, and one of the world's largest law firms, Baker McKenzie, hosted the follow-up sessions and workshops.

As a result of this event, and the unified voice of green organizations that had taken up a new and rational direction, the foundations for forest-based offset and mitigation programming for international policy development, and proposed US legislation, were being firmly established.

One year later, the International Emissions Trading Association, IETA, and a number of forest carbon companies, including ERA, co-operated at the 16th Conference of the Parties held in Cancun in 2010 to advance a renewed focus on forests, embracing REDD "reduced emissions from degradation and deforestation" (REDD and REDD+) projects for inclusion in international carbon management programming.

Five years later, at the 21st Conference of the Parties held in Paris, forests became enshrined in international climate action. Article 5, which references the original 1992 UNFCCC, positions REDD+, which has already attracted $10-billion in international investment, primarily but not entirely by governments,[105] to become a significant element of the evolving international effort to manage carbon emissions.

This shift should have implications for Canada's carbon management strategy, given that Canada's northern boreal forest represents one of the largest terrestrial carbon sinks in the world. Managing it to optimize carbon uptake and storage would appear to represent one of the country's most significant opportunities to support international carbon management and objectives.

However, Canada elected *not* to include the activity of forest management as part of its carbon accounting under the Kyoto Protocol's first commitment period, as is allowable under Article 3.4 of the Kyoto Protocol. As a consequence, while any carbon benefits derived from managing these lands would be reported under the UNFCCC, they would not be counted with respect to Canada's obligations under the UNFCCC.

The government may have taken this position to manage the uncertainty as to whether forests would prove to be a net source or a net sink in terms of its international accounting under the protocol.

The devastating fires in 2016 in the boreal forest that surrounds Fort McMurray attest to how quickly a carbon asset can turn into a carbon liability. This meant that what were arguably Canada's best carbon management assets, forests, were "out" of the equation in terms of measuring progress to meeting UNFCCC goals.

Audacity: Founding the Markit International Carbon Registry

2009–2010

I first met Helen Robinson, founder and CEO of the TZ1[106] International Carbon Registry, in Barcelona, Spain, in May 2009. We were both attending the Carbon Expo in order to connect with more people in the international carbon space and hopefully get some business done. The Expo had attracted carbon traders, project developers, standards experts and regulators from around the world.

Helen and I took a break one afternoon, hopped on an open roof bus and toured the city, absorbing its unique architecture and vitality. And we talked. I learned that Helen had a history as a senior executive with Microsoft in New Zealand. We chatted at length about the process she had led in establishing the world's most successful International Carbon Registry.

Changes were afoot, as the London-based Markit Group Limited was preparing to buy TZ1 in June and rebrand it the Markit Environmental Registry.

Based on the discussions in Barcelona and subsequent chats, the audacious story of the founding of the world's dominant International Carbon Registry has emerged.

THE STORY BEGINS

In the not-so-distant past, the small and nascent carbon markets that had emerged in various regions around the world were at great risk of failing before they got off the ground. Many skeptics challenged the science of climate change, and at the same time there was a spectrum of opportunists thinking they could make a quick buck in this new and unexplored territory of carbon. It felt like the Wild West all over again. Just substitute carbon credits for gold nuggets.

At one of Helen's first meetings in London, held at the offices of a large and credible carbon project developer, Helen, was in for a big surprise. Upon asking where the millions of carbon credits arising from a number of well-published Asian projects were located, the gentleman she was meeting opened his desk drawer and shuffled through the files, claiming, "They were in here, somewhere..."

Helen saw a problem, and an opportunity.

It was no wonder the market was being challenged and rapidly getting a bad name almost before it started! How could market participants, corporations, governments or individuals have any faith in carbon markets if they had no confidence in the authenticity of the asset they were buying or selling? How did one know that it had not been sold or traded elsewhere?

From this point of view, it was easy to understand and see the need for transparency and integrity if these new and immature carbon markets were ever to succeed. The answer? An International Carbon Registry. In Helen's words:

> *A Registry provides markets with a centralised view of transactions and proof of ownership of assets. It is a trusted facility which, whilst it is technology based, relies upon the right processes and rules to ensure adherence to appropriate legal guidelines and standards.*

It appeared there was a critical but unfilled niche in the nascent world of carbon trading.

ESTABLISHING THE TZ1 REGISTRY

This understanding of the carbon market's needs and the identification of a gaping hole in the structure led to the formation by Helen and her associates of the TZ1 Registry. The registry was to play a major role in the increased confidence in carbon markets worldwide.

While TZ1 was founded in New Zealand (located in Time Zone 1), it was obvious that it was the world's carbon markets which needed this infrastructure – not just those of Australasia. At this point the majority of market participants were to be found in Europe and North America.

So early in 2008 with a large dose of audacity we opened offices in New York and London, and focused on building customers and long-lasting relationships with the major investment banks which had strong carbon and commodity desks. The banks were significant in helping all traders, brokers, project developers and corporates recognise the importance of using the right infrastructure to ensure market integrity. These banks partnered with the TZ1 Registry helping to build a strong reputation and ultimately leading to our rapid growth to be the largest environmental registry worldwide within 6 months of our inception.

Financial markets need a balance of supply and demand – carbon markets are no different. TZ1 recognized the strong customers in Europe and North America, and at the same time believed that Brazil and China would be playing major roles on the supply side. Initially these countries would provide the bulk of the world's supply.

Surprisingly, Brazil was the country that was advancing most rapidly, and so TZ1 staff spent many months there with federal and state government officials. Meetings held with the president of Congress and the senators and ministers were critical in helping Brazil understand what initiatives were important and how environmental markets could play a significant role in their economic growth and well-being. TZ1 succeeded in building strong relationships and gained majority market share both in supply and early demand and emulated this approach across other developing country markets.

We invested heavily in the rules (the Registry Terms and Conditions). Our legal partners, Baker & McKenzie (Sydney), were outstanding. It seemed that not a day went by when we didn't uncover a new requirement in the market and needed to go back and enhance our Rules to ensure our customers and all market participants were protected, always. Managing risk is key in making markets secure – the right Rules are paramount.

Business is business – everywhere. The same elements that ensure business success anywhere applied in the launch and execution of the TZ1 Registry. In business there are a couple of key requirements in order to be successful: scale, governance *and* the right balance of flexibility and control.

Instinct, or should we say being "savvy," were probably more important in setting up the right company meeting the right business need. It was essential in the early days to remain flexible in our approach. New markets, new participants, new standards, new governments, new asset types meant being nimble, and thinking on our feet was imperative.

BUILDING SCALE

Helen posed to her team the key questions: Where were the biggest carbon market participants to be found? How could they gain market access? And how could they build a trusted brand? Setting off to create the best and biggest registry took a great deal of learning about the market. Fortunately there were an unlimited number of carbon trade shows and conferences, such as the Carbon Expo where we met (which I only now realize was by design), where the people in the know were going to build and foster strong relationships.

Hard work, long days, creative marketing, speaking and chairing events, all undertaken to generate discussion and interest. Not to mention well-attended dinners at fabulous local restaurants. Nothing like great food, wine and company to build long-lasting relationships!

Helen also understood that a key aspect to building a business is being the company people want to work with – professional, friendly and caring, working with individuals and organizations in the right and appropriate way – without compromise. Helen attributed much of the success in building a great registry business to the way in which her team believed in what they and their customers did. Helen unfailingly ensured that her team provided the best experience and the ultimate in service.

All TZ1 and Markit staff will reiterate the mantra that I continually dictated – are we being the company people want to deal with?

It was so much fun in those early Registry days, when we didn't know what we didn't know. What we did was to listen, learn and think on our feet to provide fundamental services to our clients, to always ensure our integrity was maintained and to listen and adapt to evolving market demands.

CARBON STANDARDS

Carbon standards are typically internationally recognized independent programs which provide a credible mechanism or protocol for validating methodologies and verifying carbon assertions. Helen understood that the registry, in order to be successful, would have to list the world's most accepted standards at that time, including the ISO 14064-2, VCS, the Gold, Carbon Fix, Social Carbon and Plan Vivo standards.

A critical question TZ1 faced was how the registry would choose which standards to list. As the registry was not a science-based organization it could not make any judgment on the quality of the project or the resulting assets. Initially there were key criteria, such as organizational assessment, that needed to be approved before accepting a new standard's credits.

Then for every issuance of credits, the registry would ensure that the required processes and documentation according to standards rules were met. The market itself would decide what the value of these assets should be, largely based on which organizations and standards were involved.

To address the skeptics, corporations ensure the authenticity of the offsets they use to reduce their carbon emissions (retirements of credits) by the use of an internationally recognized, credible registry. This provides certainty to the public, their stakeholders and auditors of their emissions management claims and helps manage any risk associated with offsetting.

THE ROLE OF GOVERNMENTS

Helen saw that governments play a critical role but knew that most were blissfully unaware that they were strangling the nascent market.

> *Timing is everything, and unfortunately we were right in the onset of the* GFC *[Global Financial Crisis] and many governments assumed "more important" priorities than helping the world become sustainable. In the* USA *and Canada varying states legislated carbon offset markets, and Europe led in the compliance space. Most, however, paid lip service, including Australia and New Zealand.*
>
> *Putting a price on carbon seemed too politically suicidal. In developing nations more activity transpired with governments legislating to take advantage of funding in carbon projects.*

Effective domestic programs continue to be a missing element, although Article 5 of the Paris Agreement at COP 21 provides encouragement for such programming to expand:

> *Domestic schemes must reward the right behavioural changes such as the implementation of green technology, carbon management programs, forestry management programs, renewable energy use, etc. This will attract international investment, create jobs and foster innovation.*

Helen and her team never doubted that trading was the right approach and that using market-based mechanisms would drive changes of behaviour. And they believed that there could never be a worldwide agreement wherein every country could agree to everything, and thought it was naïve to think otherwise. Helen saw that money is what makes the world go around, and therefore that financial markets would stimulate the right environmental activity.

> *With a balanced approach there is little doubt that carbon markets will be self-stimulating, creating a snowball effect resulting in what the world requires – better and more effective management of our natural resources, including a drastic reduction in emissions, improved water management, protection of endangered species and preservation of our natural reserves.*

Within this context, the registry[107] has played an essential role, providing participants the structure and integrity required to ensure the carbon market's success in addressing what is considered by many to be the most challenging issue to face humankind in the new millennium – climate change.

The Big Island of Hawaii: "Garden of the Goddess"

2010–2011

As humankind entered the third millennium the status of the world's ecosystems appeared as challenged as ever. "The Millennium Ecosystem Assessment,"[108] requested by United Nations Secretary-General Kofi Annan in 2000, engaged 1,360 experts from 95 countries. The findings, published in 2005, were that "approximately 60 per cent ... of the ecosystem services examined [were] being degraded or used unsustainably, including fresh water, capture fisheries, air and water purification, and the regulation of regional and local climate, natural hazards, and pests."

Over ten years later the situation is unlikely to have improved significantly, and more likely has deteriorated. In other words, one needn't look far to find an ecosystem that needs attention.

Consideration of degraded ecosystems and ecosystems at risk

often leads to a discussion around the loss of native forests in the Amazon rainforest, which have been cleared and burned to accommodate agriculture, impacting countless species. Alternatively, discussions have focused on the razing of forests in Indonesia, where palm oil plantations continue to decimate native biodiversity and habitat, leaving desperate primates such as orangutans, one of humankind's closest relatives,[109] homeless, hungry and doomed.[110]

ASSAULT ON THE SANDWICH ISLANDS

A region that does not often arise in such discussions is Hawaii, known as the Sandwich Islands by early explorers, which we usually associate with an idyllic tropical environment replete with turquoise waters, palm trees, beaches and the obligatory umbrella beverages.

However, the islands of Hawaii have been significantly impacted by a range of activities, including the introduction of alien species. In 1883, sugar plantation operators released the small Asian mongoose on Oahu in order to control rats that had been introduced earlier by the Polynesians as a source of protein. This was not successful, the rats being nocturnal and the mongoose being diurnal.

The mongoose subsequently decimated ground-nesting bird populations on the other islands to which it was inadvertently, or purposely, introduced. This aggressive predator now has populations established on Maui, the Big Island and Molokini, and most recently has been reported on Kauai, with similar impacts.

Agricultural operations, such as pineapple plantations[111] and cattle ranches,[112] have destroyed hundreds of thousands of acres of native forest. Resource extraction, primarily the harvesting of highly prized Ohi'a and Koa trees, which are found on humid upland slopes, has also taken its toll.

Each Hawaiian island has an arid (leeward) and a humid (windward) side, driven by the trade winds. As a consequence, the range of climatic types as one rises from sea level to volcano mountaintops is truly remarkable. For example, the island of Kauai is home to some of the highest reported rainfalls on the planet, reaching over 400 inches (just over 10 metres) per annum. This high rainfall turns the north slopes of the jutting interior mountains into veils of white over green visible from Princeville and Hanalei Bay to the north. Located only 23 miles away, on the Kekaha coast, is an arid coastline that rarely experiences precipitation of any significance whatsoever.

A number of landholding and conservation organizations operate in Hawaii, including Conservation International and the Nature Conservancy. One very large landholder is the Kamehameha Schools Trust, or Bishop Trust, named after Bernice Pauahi Bishop, the Hawaiian monarch who on her death bequeathed millions of acres to the schoolchildren of Hawaii. The trust's landholdings are vast and include a number of oceanside resorts on the major islands, and many other commercial, agricultural and forested properties.

TAKING STOCK

Spring 2008 saw the trust focusing on carbon issues that might be associated with its billions of dollars in landholdings. To secure some perspective on its carbon assets and liabilities, and to undertake a review of the carbon sequestration potentials of some of its holdings, the trust contracted with ERA Ecosystem Restoration Associates Inc., the company I had co-founded with Bart Simmons in Vancouver in 2004 (see Chapter 9).

ERA completed its review and soon thereafter began to work with the Nature Conservancy on the Big Island to determine if a Koa restoration project could be structured that would yield validated and verified carbon offsets that might finance the restoration. The understanding was that these offsets would be used to help offset a portion of the tourist industry's carbon footprint.

As most of the power production on the islands was generated from the combustion of imported bunker oil, and inter-island transport was primarily via energy-consumptive jet aircraft, the tourist industry's carbon footprint was significant. Linking ecosystem restoration to addressing this footprint seemed to make sense.

Work in Hawaii was, of course, pleasant. However, slogging through damp upland forests shrouded in a cool fog is not a walk on a bikini-studded beach. That being said, occasionally there was a sublime experience or occurrence, something magical and barely believable, that simply happened.

One such experience occurred in May 2010. Bart and I were scheduled to fly in a helicopter from Hilo on the Big Island across to the northwest coast, to survey potential project sites. When we arrived at the Hilo airport, we learned that our chopper had been seconded that morning by the seismic scientists who were now swarming the island. The Big Island was experiencing heightened levels of seismic activity that had many people concerned.

When the chopper finally arrived, we quickly loaded into the machine and lifted off heading west. Several kilometres from the airport an extraordinary scene began to unfold. Peering out the bubble windows, we could see steam and smoke rising precipitously from growing cracks in the ground surface, scorching the adjacent vegetation. Our view from the apparent safety of our machine, flying 100 metres above the turmoil, was unforgettable.

Unforgettable also was the subsequent chopper flight from a helipad farther up the island at Parker Ranch. With landholdings of over 250,000 acres, Parker Ranch had been among the largest cattle ranches in the US and was beginning to look at the potential carbon assets associated with its holdings. After the machine had warmed up, the pilot began to pull on the joystick. We began to ascend, but after only 10 to 20 seconds a bright red warning light on the dashboard of our French AStar began flashing. Apparently a sensor had detected shards of metal in the transmission oil of our chopper's turbine engine, indicating one or more gears could be unravelling. The pilot pointed to the persistent warning light, then to the ground and dropped us to the ground post-haste. Unforgettable, yes; fun, not so much.

A DRIVE WITH KAM

The most enduring magical[113] experience, this one occurring deep in the upland hills of the Big Island's west side, is the one of a handful of truly remarkable experiences in Hawaii that I will recall to the end. We had arranged to meet a conservation officer working for the Kamehameha Schools Trust we had approached regarding restoration work. Kam, a native Hawaiian male in his mid-thirties, was going to show us some potential Koa restoration sites. With Kam driving, we traversed the uplands in a three-quarter-ton 4 × 4 pickup. On the cab floor was a large, sharp machete-like knife, and in the truck's box behind us, were two hounds – just in case a wild pig was sighted.

Kam drove on for almost two hours, stopping occasionally so we could assess the various sites. We absorbed the changing scene, and continued to bump along an ever-diminishing track. I recall seeing deciduous trees that were obviously related to the salmonberry shrubs native to coastal BC. But everything about them was exaggerated. While the leaves were identically shaped, they were much larger – as were the translucent reddish-orange berries. If I didn't know better, I would say they were genetically modified. And of course they had been, by the creative forces of natural selection over time.

We continued down the track, until we came to a fence. Nobody was talking.

Kam quietly stepped out of the truck, and asked us to do the same, respectfully directing us to a spot to wait. He asked us to remain silent. On the other side of the fence was the virgin

Hawaiian forest, and something more. Kam paused for a long moment, facing the forest with elevated eyes.

Then, at first softly, and then with growing strength, he began to sing.

I was stunned. I expect that anyone who has passed by a Hawaiian church on a Sunday morning, and has heard the uniquely joyful, melodic and powerful sounds of Hawaiian singing emanating from the open doors, has been touched forever.

Kam's voice was deep, enthralling and spiritual. He sang across to the virgin forest for several minutes, going up and down scales I had never heard. Finally he went silent. Kam waited a few more minutes and then turned and looked at Bart and me and spoke quietly and reverently:

> *I asked the Goddess to see if it would be alright for you two to enter her forest. She replied that yes – you are welcome. We are going in now, please walk quietly.*

Bart and I glanced at each other, speechless. We peered past the gate into the riot of vegetation before us. We saw ferns the size of trees back home. We heard invisible birds calling in a sputtered cacophony and unseen creatures scuttling through the bushes before us. It occurred to me that Michael Crichton, who had borrowed my square tree invention for his novel's opening, was born in Hawaii, and damned if it didn't appear that we were peering right into Jurassic Park.

Bart and I fell in behind Kam. We paused. And then we stepped silently through the gate and into Laka's Garden.

Carbon from Space

2002–2017

At the time of the Apollo missions, few researchers other than Charles Keeling and his colleagues were focused on measuring planet Earth's atmospheric CO_2 levels. However, after the signing of the UNFCCC in 1992 and the treaty's ratification in 1994, when it became international law, it became increasingly apparent that scientists and policy makers would need the capacity to measure and track atmospheric CO_2 concentrations on a global scale.

This capacity would be needed to inform the development of science-based regulations, programs and actions aimed at achieving the treaty's objectives, and it would provide the means by which to measure success, or the lack thereof, in achieving the goals. This of course would require a much broader foundation of data than could be provided by the current suite of measures, primarily focused on North America, Europe and a few ocean areas.

Measures over the tropics, the boreal zone and sub-polar forests and tundra covering much of Canada and Russia, were few.

The Paris Agreement signed in December 2015 at COP 21 would confirm these needs and add further demands for atmospheric CO_2 measurement capacity: CO_2 concentrations would need to be measured at a myriad of locations, over the oceans and continents, east to west, pole to pole, season to season, around the world, for decades if not centuries.

The existing patchwork of traditional monitoring programs, involving flasks, chemical tests, sensors, observation towers and aircraft, were not going to cut it. Given the geographic and logistical challenges, might orbiting satellites be able to play a role?

MEASURING AND MONITORING ATMOSPHERIC CO_2 CONCENTRATIONS FROM SPACE

The answer to that question is an emphatic "yes," and a number of satellite systems are on the job, with more to come.

The earlier space-borne satellites, including NASA's AIRS, and TES and the European IASI instruments, were measuring CO_2 by means of sensors tuned to the thermal infrared emissions. This provided information in the lower 6–10 kilometres of the atmosphere (mid-upper troposphere). Consequently they were not of much use in identifying higher or lower concentrations close to land that would imply significant CO_2 sources and removals.

Sensors tuned to shorter infrared wavelengths that provide CO_2 data on the lower troposphere have been successfully launched by the European Space Agency in 2002 (SCIAMACHY), Japan (GOSAT) in 2009 and NASA (OCO-2) in 2014.

In December 2014, NASA reported that OCO-2 had produced its first global maps, showing averaged CO_2 concentrations from October to mid-November 2014, which are highest above northern Australia, southern Africa and eastern Brazil. The elevated levels were attributed to the burning of savannahs and forests. Scientists have implied that there may be surprises, as OCO-2 data are not always consistent with existing CO_2 data.

The priorities in the newer satellites are the capacity to measure CO_2 close to the Earth's surface, at weekly or monthly timescales. Furthermore, the ability to aim sensors to collect data from specific targets is of growing importance. This requires the spacecraft to manoeuvre.

Future needs are with respect to denser and more frequent measures that will support the growing needs of UNFCCC signatories. Also, special purpose satellites that are focused on, for example, the tropics would be highly valuable in assessing the success of REDD+ and other climate mitigation programming. Integration of satellites and ground-based measures represent another priority.

Finally, the International Space Station, now equipped with high-resolution sensors and high-definition video cameras that are aimable, could play a role in securing detailed CO_2 information around specific targets, such as a restored forest, an agricultural operation or an experimental ocean fertilization such as discussed in Chapter 7 ("Carbon and the Oceans").

MONITORING ATMOSPHERIC CO_2 UPTAKE/ REMOVALS FROM SPACE

From a carbon management perspective, the importance of being able to detect and monitor carbon removals from the atmosphere and into the land base or oceans cannot be overstated. The photosynthesis of plants[114] is the only process known, natural or technological, that removes significant volumes of CO_2 from the atmosphere. And we learned in Chapter 3 that this photosynthetic process of carbon removals can be detected and potentially measured through fluorescence, which re-emits 1–2 per cent of the sunlight received, but at longer wavelengths.[115]

As it has turned out, fluorescence can indeed be detected and measured by specially engineered space-borne sensors, tuned to the right frequencies[116] and mounted on satellites orbiting high above the Earth.

The first observations of terrestrial[117] chlorophyll fluorescence by satellite were reported in April 2007 by Luis Guanter and his science colleagues at a symposium in Montreux, Switzerland.[118] This team used a "medium resolution imaging spectrometer" (MERIS) mounted on the European Space Agency's Envisat satellite to detect fluorescence from a forested area near La Mancha, Spain, for one day only. However, this work confirmed the capacity of space-borne sensors to measure fluorescence arising from photosynthesis in forest trees, and given the strong correlation between carbon uptake and fluorescence that we and other researchers have shown, this work represented a major step forward to being able to detect, measure and monitor carbon uptake from space.

Three years later J. Joiner and his colleagues published the "First Observations of Global and Seasonal Terrestrial Chlorophyll Fluorescence from Space" in the November 2011 issue of *Biogeosciences Discussions.*[119] This work focused on several locations, including two in Amazonia, and utilized thermal and near-infrared sensors mounted on the Japanese Greenhouse Gases Observing Satellite GOSAT. Maps were produced for the months of July and December 2009 for a variety of regions in North America, Africa and Australasia.

These maps showed sharp contrasts in plant fluorescence, implying carbon uptake as well, between seasons. As would be expected, the Northern Hemisphere fluorescence production peaked during July, while in the Southern Hemisphere it peaked in December. These peaks reflected the respective season of maximum plant growth, and water stress, for each hemisphere.

NASA reported in June 2011 that its scientists had produced "ground-breaking global maps of land plant fluorescence." The maps were generated from data gathered in 2009 from a spectrometer aboard GOSAT. The first of their kind, these maps of fluorescence from land plant life showed stronger photosynthetic activity in the Northern Hemisphere in July, as would be expected from seasonal plant growth peaking at that time.

Three years later, using its "Orbiting Carbon Observatory-2" (OCO-2) mentioned earlier, NASA was able to measure and map plant fluorescence globally from August through October 2014.

Since that time, the interest and capacity for measuring carbon sequestration and fluorescence from space-borne platforms have continued to grow. In November 2015, the European Space Agency

(ESA) announced its plans to launch a satellite that would "track the health of the world's vegetation by detecting and measuring the faint glow that plants give off as they convert sunlight and the atmosphere's carbon dioxide into energy." The agency noted that its Fluorescence Explorer (FLEX) satellite will "improve our understanding of the way carbon moves between plants and the atmosphere and how photosynthesis affects the carbon and water cycles."[120]

FLEX, the agency's eighth Earth Explorer satellite, is scheduled for a launch by 2022, and a number of other systems are under design by various teams around the world.

MEASURING FOREST BIOMASS FROM SPACE

To date, radar satellites have seen limited application to the monitoring of forest biomass, which is on average ~40 per cent carbon by dry weight, because they have not been equipped to probe with the longer wavelengths required to penetrate the forest cover. However, in May 2016 came the announcement of a new European satellite equipped with "P-Band." The "Biomass Mission," which will launch on a Vega rocket in 2021, is part of ESA's Earth Explorer program. The spacecraft will be equipped with a longer-wavelength P-band sensor to provide comprehensive coverage of the global forest.

The Biomass Mission's novel space radar will make 3D maps of forests, to improve our understanding of how carbon is cycled through the biosphere. The data generated will be of critical importance to climate research and will support the establishment of biomass baselines and monitor changes to the status of global forest resources. The new spacecraft is to be assembled by the UK arm of Airbus Defence and Space.

MEASURED BY THE SLICE

The new P-band radar, pulsing with a wavelength of 70 centimetres, is a type of instrument that only recently has begun to be employed in orbit. At this wavelength, the radar can look through the leafed canopy of forests to the woody biomass below.

This new approach will scan slices through the trees on repeat passes to build up a picture of how much woody material is present. Global maps should be produced every six months for at least five years.

In the words of Professor Shaun Quegan, who was one of the key proposers of the mission:

> *Effectively, we'll be weighing the forests.... We'll know their weight and their height at a scale of 200m, and we'll see how they are changing over time.*
>
> *This will give us unprecedented information on deforestation – on how much carbon is going into the atmosphere from this source. At the same time, we'll also see how much carbon is being taken up in regrowth.*[121]

At the same time, Canadian-based UrtheCast Corp. was planning to develop a 16-satellite constellation that combines an advanced radar with two simultaneous X and L bands that will capture both the forest canopy and the ground with trailing multi-spectral optical. This will allow a much more detailed look at specific problem areas. Forestry scientists working in conjunction with the broad cover P-band European satellite and Urthecast's multiple-sensor targeted satellite constellation will acquire the best of both systems to provide unprecedented biomass and forest health information.

Given these systems, with more innovation to come, the breadth of information that will be available to future resource managers is staggering. But such is the task at hand, and we have only just begun in earnest.

Return to the Salish Sea

2014–2016

It was late October and a light drizzle was falling on the waters off the forested coastline of Howe Sound, BC. A boating friend and myself had ventured out for a few hours from our safe moorage at Fisherman's Cove to take in one of the last days of a seemingly endless Indian summer. With the engine silent and the radio off, it felt like a good time to pause – a time for contemplation and perhaps a warm beverage.

Howe Sound is a triangular-shaped archipelago, stretching from Bowen Island at its southern extremity, 42 kilometres northeast to the town of Squamish, or "Sḵwx̱wú7mesh"[122] in the Coast Salish language. The city of Vancouver sits 25 kilometres southwest. Howe Sound is located within a complex and much larger network of waterways known as the Salish Sea.[123] British Columbia's Desolation Sound defines the northwest border of the Salish Sea, while

Washington State's Hammersley Inlet located at the head of Puget Sound marks the southern border

Only a few kilometres north of here, off Bowen Island's Point Finlayson, as a boy of 8, I had seen my first bald eagle. I was thrilled, and grabbed my father's arm as I pointed to the sky. To me, discovering this amazing bird of prey in our "own backyard" was far more valuable than stumbling on a nugget of gold. A few years later, I was stunned to discover a California alligator lizard on nearby Keats Island. Lizards residing on British Columbia's raincoast? It seemed unfathomable, but there "it" was.[124]

As the tide and breeze took us slowly toward nearby Snug Cove on Bowen Island, a growing chill drove me into the boat's small cabin. From there, as the wind freshened over the sound, I peered through the windows.

After several moments of gazing over the choppy waters, I was startled to see what appeared to be a vertical water column, perhaps three metres high. Then it was gone. I said nothing, thinking that my eyes were playing tricks. But there it was again, perhaps ten metres off our starboard bow – but this time I saw the shadow of a large horizontal body beneath the column of water and mist.

A CHILDHOOD DREAM

The obvious dawned on me: just beneath the ocean's surface was a large whale, a Pacific humpback, as it turned out. I was astounded. While I had secretly hoped to see whales in Howe Sound since I was a child, it had never happened – until now. I was both flabbergasted and elated.

The whale's departure was signalled by its huge tail fluke pointed skyward just before the leviathan disappeared on a deep dive, presumably to resume feeding on the biologically rich waters.

My boating friend and I returned to our moorage saying very little, but smiling.

A few evenings later I shared the story with some boating friends, musicians and the proprietor of Hugo's, our community's live music venue in Fisherman's Cove where our boats were moored. News of the sighting spread quickly, and others began posting their sightings on YouTube.[125] To my surprise, it appeared there was more than one whale. What was going on? Humpback whales in Howe Sound?

Then it dawned on me that there were other signs of change in the sea where I had grown up.

Earlier that autumn I had been drawn from my home, about 8 kilometres to the northeast of the whale sighting, by the spectacle of dozens of marine mammals cavorting and frothing the waters 50–100 metres offshore. After a short drive downslope and a quick walk to a rocky point lookout for a closer look, I saw a pod of approximately 70 white-sided dolphins. Moments after my arrival, the pod turned and raced to where I stood – jaw hanging. The dolphins swished past me at high speed, literally slicing through the water.

This was another absolute first-time experience for me, and another dream had come true. Many times I thought I had seen dolphins in the sound, but this was always the product of wishful thinking mixed with the dancing wake of a boat unseen. But this sighting? This was the real deal.

The whale encounters continued, as did those with dolphins. As well, transient orcas or "killer" whales that normally reside off the outer coast of Vancouver Island were now making 150-kilometre excursions to Howe Sound.

What was happening?

SOME HISTORY[126]

In 1791, when the crew of a Spanish ship became the first Europeans to set their eyes upon this same sound, and the steep forest-covered islands, fjords and mainland mountains that frame it, the sea was literally teeming with marine life. The Spanish captain, Francisco de Eliza, named it "Boca del Carmelo," presumably in reference to the biblical Mount Carmel in Israel.

One year later a British ship under the command of Captain George Vancouver arrived, and one of his officers offered up the name "Howe Sound" in respect to Admiral James Howe. This name has endured.

Captain Vancouver no doubt encountered the Coast Salish First Nations who had occupied the islands, mainland bays and uplands of this region for thousands of years. These First Nations lived in an abundance of rich seafood. They were surrounded by waters teeming with migrating runs of salmon numbering in the many millions of fish per year.

This extraordinary bounty of salmon, along with herring, anchovies, cod, rockfish and many other species of fish and invertebrates,[127] attracted Pacific humpback, grey, minke, right and sperm whales. Orcas, white-sided dolphins, harbour seals, harbour porpoises, Dall's porpoises and Steller's sea lions also plied the waterways, while river otters foraged in the estuaries seeking fish, crustaceans and amphibians.

Millions of resident and migrating seabirds, including cormorants, scoters, swans, ducks and geese, occupied and foraged the estuaries and shorelines as well. During the salmon spawning season, an estimated 35,000 bald eagles congregated on trees and sandbars along the rivers. On the ocean shores and along the rivers, grizzly and black bears, as well as wolves, coyotes, raccoons and mink, joined in the feast.

This ocean bounty was complemented on land by populations of blacktail deer and Roosevelt elk, and the area was graced by the mild winter climate that characterizes the coastal region. This abundance attracted small groups of adventuresome Europeans who settled, worked on their homesteads, and began to build their new lives in the area, living in relative harmony with Nature.

The early phase of the area's settlement was built entirely upon the area's biological bounty. Lands rich with fur-bearing mammals drew trappers, while countless runs of migrating salmon headed to nearby spawning rivers drew fishermen, and the region's unique endowment of forests attracted loggers from around the world. The regenerative capacity of the ecosystems had not been exceeded, and the Europeans and First Nations lived side by side sharing the natural resources of their new and traditional homes.

However, this balance was to be disrupted. It began with the discovery of gold in the Squamish River and areas to the north, which attracted throngs of fortune seekers. While the impacts of this discovery and the influx it catalyzed were transitory, the discovery and exploitation of other minerals in the nearby mountains that framed the sound would disrupt the ecology of the area for a very long time, perhaps forever.

This second wave of disruption was triggered in 1890 with the discovery of an enormous body of copper ore by a medical doctor and his guide who were deer hunting on Britannia Mountain, which is located at the northeast extremity of the Salish Sea.

A claim was staked, the ore body assessed, and in just over a decade, the Britannia Mine was built. During its active life it would become the largest copper mine to operate in the British Empire.

Over 60,000 people lived in the mining community over the 70 years of operations, during which 50 million tons of ore were extracted. If a train was assembled to accommodate this production, it would stretch 4000 kilometres, across the continent from Vancouver to Quebec City. At its peak of operation in 1929, the mine produced 17 per cent of the world's copper.

The Britannia Mine permanently ceased operations in 1974 when economically viable ore reserves ran out. However, the pollution continued, with over a ton of copper and zinc being discharged every three days into the waters of Howe Sound.

The mine operations had changed the ecology of the receiving waters of Howe Sound dramatically, and would continue to do so for another three decades after closing. Visible to those who chose to look beneath the surface was a virtual "dead zone" where Britannia Creek, which carried toxic effluent from the mine's operations, entered the ocean. A Fisheries report revealed that chinook salmon held in bioassay cages at the mouth of Britannia Creek died in less than 48 hours. The entire range of effects on marine life in the sound can only be guessed at now, but they were significant. The question was, were the impacts reversible?

Following a sustained public awareness campaign led by the internationally renowned International Rivers Day advocate Mark Angelo during the 1980s and 1990s, good intentions started to become reality. The toxic flow from the Britannia Mine site was finally addressed in 2006 after the provincial government and industry joined forces to finance the construction and operation of a treatment system that would be handling 4.2 billion litres of contaminated runoff per year, removing 226,000 kilograms of heavy metal contaminants. With a capital cost approaching $60-million and an operating cost of $1-million per year, the treatment system will operate for many decades, perhaps centuries, to come.

In concert with this treatment facility, halfway up the mountain University of British Columbia engineers and a Squamish company designed and installed a concrete plug in order to contain the pollution from an upstream tributary of Britannia Creek.

The collective benefits of the treatment facility and the concrete plug, which brought the site into compliance under the British Columbia's Contaminated Sites Regulations, were undeniable and virtually immediate. Over 30 years after the mine closed, the heavy metal pollution of the sound had been reduced by 99 per cent. Local residents reported that marine life began to return to the once "dead zone" almost immediately.

The full impacts and effects of Britannia, Woodfibre pulp mill and other industrial activities in or near the sound on marine mammals, fish and other marine life, will never be completely known as there is no pre-industry baseline to compare against.

However, and specifically relating to whales, it is likely that the near-disappearance of these large mammals from Howe Sound in the last century was at least in part the result of commercial whaling operations. During the period of commercial whaling, Pacific humpback whale populations on the British Columbia coast plummeted from approximately 20,000 to 1,400, when a whaling moratorium was finally enacted in 1967.

Several other whale species were hunted before the moratorium. The last right whale to be seen on the coast was harpooned and killed in 1951 by whalers operating out of Vancouver and seeking other prey.

Thankfully the timeless gift of Nature – including the area's glaciers, rainforests and rivers – would eventually restore the water quality of Howe Sound. In concert with Nature, the sustained advocacy efforts, the enforcement of effective environmental protection regulations and the financing and construction of an effective water treatment system at the Britannia mine site were all essential to get the job done.

THE TIDE TURNS

Experiences similar to ours, with sightings of Pacific humpback whales, white-sided dolphins and increasingly frequent orca visitations, are being enjoyed by a growing number of residents and visitors from around the world. Remarkably, some of these sightings are occurring within the Vancouver harbour. That which was unheard of when I was growing up is becoming almost commonplace.

After 100 years of unintentional assault, the tide is clearly turning for this spectacular corner of the Salish Sea.

Epilogue

Carbon in Balance

2016–2017

KO SAMET

I blinked to consciousness in an instant, but I had absolutely no idea where I was. My eyes still closed, I could feel a strong ocean breeze and hear the staccato chorus of cicadas, chirping lizards, and the crashing of ocean waves meeting their fate on a beach. I opened my eyes, and I got it. The giant rustling leaves of banana plants, and the warm glowing lights within cottages.

Yes, I was on Ko Samet.

I was staying for a few days at a small secluded resort on the south tip of this 13-square-kilometre island, located in the northern Gulf of Thailand. It was an evening in late January, perhaps 10:00 p.m. Apparently, sitting on the deck of my cottage I had nodded off while taking in the sights, smells, and sounds of this little piece of paradise that I had stumbled upon.

There was something very different about this place, and it had to do with sounds. Beyond the natural background sounds of the elements and a fecund tropical biology doing its thing, there were no human sounds discernible – no beachside elevator music, no live music or karaoke (I had just come from Pattaya, party central).

With one exception.

There was a barely audible rumble, a sound first detected and felt by the body, and only thereafter heard by the ears. In fact, for some other folks I spoke to, there was nothing to be heard.

I am accustomed to hearing, where I live on the Salish Sea, the sounds of large ferry boats at a distance. This was similar, although it felt and sounded bigger. I recalled noticing during the day a fully loaded tanker far off on the horizon, headed out southwest. And I had seen from a taxi a few days before an enormous refinery on the nearby mainland coast. Perhaps it had supplied liquid hydrocarbons to the tanker I had seen.

Whether it was the refinery, or one or more out-of-sight tankers, or some combination thereof, mattered not. What I was hearing were the sounds of the hydrocarbon beast, and it occurred to me with great clarity that evening that it wouldn't be going away anytime soon.

Earlier that day I'd had an interesting discussion with a well-educated Indonesian businessman with close ties to the president's office, and he explained Indonesia's policy around its hydrocarbon resources. I was surprised to discover that Indonesia is the second largest coal exporter on the planet, representing close to 21 per cent of global exports. The message I heard was that the president was not going to burden his country with the incremental energy costs associated with alternative, presumably more expensive energy sources, i.e., renewables.

I recalled my discussions with officials in China 20 years ago.

The message at that time was that China has 1,000 years of coal reserves, at least, and that the country was going to use them. This two-decade-old message was echoed by recent reports of China's predicted 19 per cent increase in coal power generation power capacity over the next five years.[128] China currently burns four times more coal than any other nation.[129]

And so it becomes abundantly apparent that notwithstanding the rapid rise in renewable capacity in China and around the world, the hydrocarbon beast will not be going away soon – certainly not within the lifetime of anyone reading this account.

I had returned to Thailand to finish this book, thinking a trip and final comments from Chiang Mai, 28 years from when the story began, would "join the dots," or "close the loop" – choose a metaphor – but inside me I knew that this story and its telling were about to complete.

Home awaits, with more carbon work as another wave of "calls to action" coming into play. However, this time it is going to be different. Much has changed since we began this journey, and there is no going back.

I remember standing at the threshold of the Hawaiian Goddess's garden with Bart, Kam and the two hounds. But there was a fundamental difference. To the gateway that Humanity stands before today, there will be no return. We have heard no melodic singing, nor have we been apprised that the blessing of a Goddess will be with us as we step into the abyss.

Standing before the threshold I am reminded of the lyrics from a symphonic rock and roll band from the 1970s:

We are going, to keep growing, wait and see ...[130]

The bleak reality we must address is that the global population continues to grow unabated, and with it the demand for all natural resources. And while the development and consumption of low- or no-carbon renewable energy grows dramatically, the overall demand for energy is such that the consumption of virtually all energy resources, including hydrocarbons, will continue to grow for the foreseeable future. And so long as the resources are available, and the infrastructures are in place to deliver them, and the price is right, these resources will be consumed.

The final chapter of this book presented a "happy ending" for the Salish Sea and its fresh start as a recovering and functional ecosystem. But how is this relevant to carbon? One could accurately state that as carbon is the "mineral of life," any story around biology and natural history is automatically a story about carbon. But that would miss the point.

It is the "question of balance" that I believe to be absolutely critical at this juncture.

In December 2015 the 21st Conference of the Parties (COP 21) held in Paris brought together the largest gathering of world leaders in history to address climate change. Canada, whose emissions represent 1.67 per cent of global anthropogenic carbon releases to the atmosphere, sent well over 200 representatives to attend the conference. No other country, even those with over ten times the carbon emissions, came even close to Canada in terms of representation.

Against all odds, given the repeated shortcomings of previous COPs, COP 21 appeared to have made significant progress in securing concerted international support and co-operation to address climate change. With the establishment of the "Paris Agreement," the focus on carbon management had attained an unprecedented international profile, and as such was received with unbridled jubilance by the environmental community.

For a veteran carbon explorer who had been glued to the climate and carbon issues since well before the UNFCCC was signed in 1992, this international profiling should have been exciting.

But for me it was not at all exciting. To the contrary I felt anxious. I felt as if I had been here before.

At COP 3 held in Kyoto Japan in 1997, the failure in achieving the foundational 1992 UNFCCC goals was abundantly clear, as the original reduction goals of the convention were being largely abandoned by some of the nations that mattered, including the largest CO_2 emitter at the time, the United States.[131] Being a "developing country," China was not obligated under the Kyoto Protocol to identify emissions reduction targets. Within that context, the "Kyoto Protocol" was developed, and it appeared to bring some of the much-needed structure, including binding targets and timetables, that would be required if the convention was to succeed.

Alternatively, a skeptic might say that the Kyoto Protocol simply pushed out the timing of delivering CO_2 emissions reductions to a "commitment period" of 2008–2012, which meant another 15 years of wriggle room and no enforceable consequences for non-compliance.

The reality of the first commitment period ending in 2012 was that global anthropogenic CO_2 emissions, which were to have been reduced to 1990 levels under the 1992 UNFCCC, had ballooned to approximately 32 billion tonnes of CO_2, a 48 per cent rise from 1992.[132]

Furthermore, positions were changing. By the year 2015, China had eclipsed the US as the number one emitter at 28.03 per cent of total global emissions, with the US at 15.90 per cent. Canada's share of global emissions, as previously mentioned, was 1.67 per cent.

From this perspective, I will leave it to the reader to look at the Paris experience optimistically – but cautiously. Beyond the potential for another failure, what was nagging and tugging at me from COP 21, as well as other climate-related programming, is the potential risk of unpredicted and undesirable outcomes arising from an overly singular focus on carbon management, and the potential exclusion of other environmental and humanitarian imperatives. I had good reason to be concerned.

CARBON AND PERVERSE OUTCOMES

What might be considered an undesired outcome from a singular carbon focus with respect to direct impacts on people? I learned of one, to my surprise and dismay, during a live interview on national CBC Radio in 2010. At the time an African subsidiary of ERA Ecosystem Restoration Associates Inc. was developing forest ecosystem restoration projects in the Mai Ndombe region of the Democratic Republic of Congo (DRC). The undertakings were to be financed by the carbon offset values as "Reduced Degradation and Deforestation" (REDD+) projects that would contribute to preserving the second largest rainforest in the world.

The projects were community-focused and intended to bring multiple ancillary benefits with respect to education, health and employment. Frankly, I was proud to be associated with the projects, and pleased with the opportunity to discuss them with a national radio audience. The interview was conducted early one morning at CBC's Vancouver studios for inclusion in a popular program, *The Current*.

After some friendly off-line banter in the studio "green room," the live interview began. The host was in Toronto as I recall, so I was looking at no more than some flashing lights and a microphone. I was immediately blindsided. Rather than query me about our proposed projects in Africa, the interviewer began by referencing another organization's forest carbon project, this one in Burundi. Apparently the Indigenous peoples had been banished from the very forests that represented their homes and had provided them food, fuel and shelter for millennia. I was flabbergasted.

It was not our project, but the well-prepared host wasted no time in asking me if that is what we were "up to" in Mai Ndombe. There was an edge to the questions. It was sink or swim, and I was in my birthday suit, uprepared for pointed questioning. I methodically described our proposed projects, including the community engagement, employment and other benefits that would arise from them. But the reality was that we were very early in our programming, and primarily focused on securing agreements with communities and senior government agencies.

I had no first-hand knowledge of the project in Burundi, but if it was as described, one might see the alienation of Indigenous peoples from their forest homes as a perverse outcome.

Within this context, and returning to COP 21, it was encouraging to see in Article 5.2 of the Paris Agreement[133] the following:

> *Parties are encouraged to take action to implement ... policy approaches and positive incentives for activities relating to reducing emissions from deforestation and forest degradation, and the role of conservation, sustainable management of forests and enhancement of forest carbon stocks in developing countries; and alternative policy approaches, such as joint mitigation and adaptation approaches for the integral and sustainable management of forests,* ***while reaffirming the importance of incentivizing, as appropriate, non-carbon benefits associated with such approaches****. [Emphasis added.]*

Of course the devil is in the details, and the realities of implementation, but the contents of Article 5.2, a modicum of rationality and a great deal of hard work will be required if COP 21 is to deliver a balance of carbon and ancillary benefits as is hoped.

COP 22, held in Marrakesh in November 2016, which was intended to tie up some loose ends and turn the Paris Agreement into a blueprint for action, was very much focused on process and mechanisms, and defining key issues to be addressed.

Unlike COP 21, which enjoyed unprecedented media attention, COP 22 was overshadowed by the US elections. Concerns for the impacts of a Republican win on the Paris Agreement were few and were muted with the expectation of a Democratic win, perhaps a landslide.

The election's actual outcome brought a whole new set of

uncertainties, with the US President-elect affirming the country's commitment to full implementation of the Paris Agreement, while ordering the Department of Energy to provide a list of names of staffers who worked on climate change issues. (The Department of Energy said it would not co-operate.)[134] In June 2017, President Trump announced his intentions to pull the US out of the Paris Accord.

Even if China is now the largest carbon emitter on the planet, the critical importance of US involvement in the implementation of a successful international climate policy cannot be overstated. And notwithstanding a range of process initiatives (e.g., a fair rulebook and achieving a greater balance between adaptation and mitigation, etc.), and financial promises made or confirmed (e.g., $100-billion per year by 2020) at Marrakesh, the future of this unprecedented adventure in international policy development and implementation is highly uncertain.

EMERGING IMPERATIVES

On a bright note, as an issue and critical resource requiring management, "water" achieved a new level of focus at COP 22, with one full day devoted to the issue, and how action on water could contribute to the Paris Agreement. "Conserving, restoring and managing forests is essential to meeting global sustainable development goals, including combating desertification and water security," commented High Commissioner for Water, Forests and the Fight against Desertification of Morocco Abdeladim Lhafi, co-organizer of the event.

If it turns out that through international co-operation on addressing the climate change issue, the other critical challenges facing humankind with respect to forest and ecosystem protection and restoration and water conservation, all of which are linked, are illuminated and acted upon, the sought-for but elusive balance may yet get its day in the sun.

But what about the oceans?

ANOTHER COLOSSAL BLUNDER?

While COP 22 played out, on the 20th anniversary of the Kyoto Protocol's "colossal blunder," the most significant player on the climate and carbon mosaic, the global ocean, was lapping against the borders of 183 coastal nations. At the same time, an ever-growing number of orbiting and water-borne constellations and sensors were gathering an overwhelming bank of data and evidence on the flows of carbon to and from the biosphere's largest (by an order of magnitude) carbon reservoir, the global ocean.

But was the international community aware or attentive, as the bureaucratic processes and mechanisms were hammered out at COP 22? Was this déjà vu all over again? Perhaps not.

The COP's first Oceans Action Day was convened on November 12, 2016, on the periphery of COP 22. Princess Lalla Hasna of Morocco and Prince Albert II of Monaco opened a full-day program that illuminated initiatives to develop socio-economic opportunities from marine resources in coastal communities. International marine areas protection was also addressed.

Plenary sessions were held on "Oceans and Climate"; and

"Oceans and Climate: Science Solutions," where solutions to the core issues such as food security, mitigation, adaptation and building resilience, were discussed. At these sessions experts elaborated on some of the policies, initiatives and commitments required to achieve the Sustainable Development Goals of the Paris Agreement.

A report, *Toward a Strategic Action Roadmap on Oceans and Climate: 2016 to 2021*,[135] was prepared, which explicitly identified, in its Mitigation section, the need for "considering" ocean-based carbon capture and storage, and the need to "integrate the management of coastal carbon ecosystems ... into the policy and financing processes of the UNFCCC."

As I recall my excitement on the signing of the UNFCCC in Rio de Janeiro in 1992, and remember discussing ocean carbon on the morning of September 11, 2011 with my ASLO colleague Jonathan Phinney in Washington, it becomes apparent that the wheels of international policy development and implementation move exceedingly slow.

But the wheels can and do turn.

Perhaps the most important and valuable challenge of climate change and carbon management is that of holding humankind's feet to the fire, in terms of revealing how well we understand how the world works. Climate change and carbon have driven a rethink of virtually every discipline: from finance to forestry; from ethics to photo-physiology; from health to engineering; from biotechnology to waste management; from agriculture to oceanography; and yes, from aerospace to women's issues.

CLEAR VISION

As a university student I worked part-time as an optical technician, sizing, fitting and repairing spectacles for patients at my father's optometrist office. My father's optical prescriptions provided the miracle of clear vision. Perhaps, then, it is no coincidence I see that the climate change issue and carbon management challenges have collectively presented mankind with a new set of spectacles, or at least, the impetus to fashion them.

It is my hope that as humankind imparts its not-insignificant impacts on Spaceship Earth, we employ our new-found vision and perspective constructively and joyfully.

R.W.F.

2017

Glossary, Acronyms and Abbreviations

Algae. Algae are microscopic unicellular or multicellular organisms, that occur in fresh or salt water or moist environments. Algae have chlorophyll and are photosynthetically active, removing CO_2 from the water or air. Examples are the phytoplankton of oceans and fresh water bodies.

Association for the Sciences of Limnology and Oceanography (ASLO). ASLO was established in 1947 and was formerly known as the American Society of Limnology and Oceanography. ASLO is a scientific society with over 4,000 members and has identified as its primary the goal of advancing the sciences of Limnology (lake biology) and Oceanography.

Bioassay. A bioassay is a setup that involves the use of live animals, including finfish and invertebrates, or plants, to test the biological effects or impacts of a substance, such as a drug, hormone or toxin.

Biomass. Biomass is the mass of organic matter, usually plant matter such as forest product waste, that can be converted to fuel and used to produce heat and/or electric power. Biomass is usually measured by weight (kilograms) or volume (square metres).

Cambium. The cambium is an embryonic tissue layer of a plant that facilitates plant growth. In a tree, it takes a cylindrical geometry, forming xylem cells on its inside, which become wood, and phloem cells on its exterior, which become bark. The carbon required by the cambium for the production of tissue is derived from carbon removed from the atmosphere by means of photosynthesis.

Canadian Association of Petroleum Producers (CAPP). CAPP, with its head office in Calgary, Alberta, is a well-established industry association that represents the upstream Canadian oil and natural gas industry. CAPP's members produce an estimated 90 per cent of Canada's natural gas and crude oil. CAPP is an active lobbyist, developing and presenting the interests and positions of the industry with respect to environmental regulations and climate change policy.

Canadian Energy Pipeline Association (CEPA). CEPA was established in 1992 to represent the Canadian pipeline industry's interests. CEPA's current membership of transmission pipeline companies represents 97 per cent of Canada's daily natural gas and onshore crude oil delivery to North American markets.

Canadian Gas Association (CGA). The CGA was founded in 1907 to be the voice of Canada's natural gas distribution industry. CGA members are natural gas distribution companies, transmission companies, equipment manufacturers and other service providers. CGA members deliver natural gas to approximately 6.5 million homes and institutions, reflecting approximately one-third of all the energy needs in Canada.

Cantor Fitzgerald, LP. Cantor Fitzgerald is a financial services firm that was founded in 1945. The company has over 5,000 institutional clients, serving the market with investment banking, prime brokerage and commercial real estate financing. The firm is also active in new businesses, including advisory and asset management services, gaming technology, e-commerce and other ventures. Through its subsidiary CantorCO2E, the firm has played a leadership role in carbon credit trading and asset management. In 2011 CantorCO2E was acquired by BGC Partners, LP, and became BGC Environmental Brokerage Services, LP.

Carbon. In scientific terms, carbon is the non-metallic chemical element having the atomic number 6. In the vernacular of climate change and carbon management, e.g., "putting a price on carbon," or the "low-carbon economy," the word carbon is used broadly, referring to carbon dioxide or other gaseous carbon compounds such as methane.

Carbon Credit. A "carbon credit" is a carbon offset that meets a specific standard that qualifies it to be used by industrial emitters to address governmental emissions reductions policies, regulations and requirements and/or to be fungible (tradable) in an emission reduction trading system, in an identified jurisdiction.

Carbon Dioxide (CO_2). Carbon dioxide is a colourless gas that is present in the Earth's atmosphere at a current concentration of approximately 400 ppm. Carbon Dioxide is the reference greenhouse gas, having a global warming potential of 1.

Carbon Expo. The Carbon Expo is an annual trade fair directed to advancing the future of carbon markets, finance and trade. The Carbon Expo attracts over 2,000 participants and 200 expert speakers from over 100 countries, including representatives of governments, NGOs, companies, financial institutions and research institutes.

Carbon Neutral Government Regulation of British Columbia. BC's Carbon Neutral Government Regulation requires all public sector organizations in BC to: (a) reduce emissions as much as possible each year; (b) measure any remaining GHG emissions arising from buildings, vehicle fleets, paper use and travel; (c) purchase an equivalent amount of emission reductions (offsets) to achieve zero net emissions; and (d) report on their achievements.

Carbon Offset. A carbon offset is a quantified reduction, avoidance or removal of emissions of carbon dioxide or another greenhouse gas, made in order to compensate for or to counterbalance an emission made elsewhere. Carbon offsets may be used to address either regulated or voluntary emissions reductions, obligations and objectives.

Carbon Sequestration. A natural or artificial process by which carbon dioxide is removed from the atmosphere or a waste stream and secured in solid or liquid form in a reservoir other than the atmosphere.

Chicago Climate Exchange (CCX). The Chicago Climate Exchange (CCX) was established in 2003 by futures guru Richard Sandor to be North America's first voluntary, legally binding greenhouse gas trading system for offset projects. CCX ceased operations in 2010 due to the very limited volume of carbon offset trading in the US.

Chlorofluorocarbons (CFCs). CFCs are a class of compounds including carbon, hydrogen, chlorine and fluorine, typically gases used in refrigerants and aerosol propellants. They are harmful to the ozone layer in the Earth's atmosphere because of the release of chlorine atoms upon exposure to ultraviolet radiation

Clean Development Mechanism (CDM). The Clean Development Mechanism is one of the mechanisms defined in the Kyoto Protocol.

CDM was directed to meeting two objectives: (a) to assist parties not included in Annex 1 in achieving sustainable development and in contributing to the ultimate objective of the UNFCCC, and (b) to assist parties included in Annex I in achieving compliance with their quantified emission limitation and reduction commitments.

Climate Change. Climate change is a statistically significant change in weather patterns as defined by precipitation, temperature and extreme weather events that last for extended periods of time (i.e., decades to millions of years). Climate change is believed to be caused by a combination of biological processes, variations in solar radiation received by Earth, and plate tectonics and volcanism, as well as human activities, including industrial CO_2 emissions, deforestation and land use change.

Climate Change Adaptation. Climate change adaptation refers to actions taken to address climate change impacts that are either current or inevitable. This could involve preparing storm water infrastructure to prepare for extreme weather events or adjusting planting of forests and agricultural crops to accommodate changing climatic conditions.

Climate Change Mitigation. Climate change mitigation refers to actions taken to reduce and stabilize the releases of heat-trapping greenhouse gases in the atmosphere. Such actions could include switching away from high carbon-content fuels (e.g., switching power plant fuel from coal to natural gas), deploying renewable energy to replace fossil fuel-based power generation, and the use of forests and agricultural operations to sequester and store more carbon from the atmosphere.

Climate Stewardship Acts. A series of three acts introduced to the United States Senate by Senator John McCain, Senator Joseph

Lieberman and a number of other supporters over the period 2003 to 2007. The Acts proposed the introduction of a mandatory cap and trade system for principal greenhouse gases. The three acts failed to gain enough votes to pass through the Senate.

Community Ecosystem Restoration Program (CERP). CERP is a large-scale urban ecosystem restoration program that began in the District of Maple Ridge in 2005 and expanded to include projects in the District of Mission, the City and Township of Langley and Metro Vancouver. CERP has generated over 1,600,000 tonnes of carbon offsets that have been registered on the international Markit Registry for international voluntary carbon offset markets.

Compliance Carbon Market. Compliance markets are established for the trading of carbon offsets that are fungible within a jurisdiction and needed to comply with a regulation. The Compliance Market is a market for carbon offsets created by the need to comply with an emission reduction regulation. In the case of a cap and trade emission reductions market, emitters buy and sell carbon offsets to comply with the specific cap or limit imposed on their emissions within a jurisdiction.

Conference of the Parties (COP). The COP is the supreme decision-making body of the United Nations Framework Convention on Climate Change (UNFCCC) that was established in 1992. The COP is comprised of the states party to the convention, and adopts and makes decisions necessary to promote its effective implementation, including institutional and administrative arrangements. The COPs are held annually and numbered in terms of their sequence. The most recent COP (COP 22) was held in Marrakesh in November 2016.

CO_2e. The standard measurement unit (meaning CO_2 equivalent) used in carbon accounting and trading. It is used to express the impact per tonne of non-CO_2 greenhouse gas releases such as CH_4 and N_2O emissions, in terms of the number of tonnes of CO_2 released to the atmosphere that would result in the same amount of radiative forcing.

David Suzuki Foundation (DSF). DSF is an environmental not-for-profit advocacy organization that, like Greenpeace, had its beginnings in Vancouver. The foundation's original goal was "to find and communicate practical ways balancing human needs with the Earth's ability to sustain all life." The foundation is supported by approximately 30,000 donors.

Electric Power Research Institute Inc. (EPRI). EPRI represents a consortium of energy companies and conducts research and development relating to the generation, delivery and use of electricity on their behalf and for the public. EPRI is an independent, non-profit organization that engages scientists and engineers as well as experts from academia and the industry to help address a range of challenges and opportunities in the provision of safe, reliable, affordable and environmentally responsible electricity to society.

Environment Canada. Environment Canada is the ministry of the Government of Canada having lead responsibility for coordinating environmental policies and programs as well as preserving and enhancing the natural environment and renewable resources. The ministry was renamed following the 2016 federal election to ***Environment and Climate Change Canada***, likely to demonstrate Canada's renewed participation in the international efforts to address climate change.

Environmental Management System (EMS). An EMS refers to a comprehensive, systematic, planned and documented structure that integrates the organization, planning and resources for implementing, monitoring and reporting to ensure compliance and protection of the environment. The EMS must address governance and reporting to ensure the board of directors is made aware of significant environmental issues in a timely manner.

Envisat. Envisat was a satellite that was launched in 2002 with ten instruments aboard and was the largest civilian Earth observation mission to date. Envisat carried two atmospheric sensors for monitoring trace gases. It also carried advanced imaging radar, radar altimeter and radiometer instruments and a medium-resolution spectrometer that featured ocean colour. Following a loss of connection, the Envisat mission ended on April 8, 2012.

European Union Emissions Trading System (EU ETS). The European Union Emissions Trading System was established to address emissions goals of the UNFCCC and was the first large greenhouse gas emission trading scheme in the world. Launched in 2005, it covers over 11,000 power stations and industrial plants in 30 countries, whose collective carbon emissions represent approximately 50 per cent of Europe's total emissions.

Fluorescence. The visible and invisible radiation emitted by certain substances, including the chlorophyll pigment of plants, when excited by incident radiation of a shorter wavelength. In the case of the chlorophyll pigment of plants, the fluorescence is in the near-red and infrared wavelengths.

GLOBE Foundation of Canada. The GLOBE Foundation of Canada is a Vancouver-based, not-for-profit organization dedicated to finding practical business-oriented solutions to the world's environmental problems. Founded by Dr. John Wiebe in 1993, the foundation advances and promotes environmental business opportunities through the GLOBE Conference Series, research and consulting, project management, communications and awards programs.

Global Warming Potential (GWP). GWP is a measure that was developed to quantify the global warming impacts of different greenhouse gases in comparison to CO_2, which has a GWP of 1. A GWP is calculated over specific time intervals, typically 20, 100 or 500 years.

Greenhouse effect. The term used to describe the heating of the atmosphere owing to the presence of water vapour, carbon dioxide, methane and other gases, that allow incoming sunlight to pass through but trap a portion of the longer wavelength energy emitted back from the planet's surface by thermal radiation, as heat.

Greenhouse Emissions Management Consortium (GEMCo). A not-for-profit corporation formed in 1994 by a group of Canadian energy companies in order to demonstrate leadership in, and develop capacity for, voluntary and market-based approaches to greenhouse gas emissions management.

Greenhouse gases (GHGs). The atmospheric gases, principally water vapour (H_2O), carbon dioxide (CO_2), methane (CH_4) and nitrous oxide (N_2O), that contribute to the greenhouse effect by absorbing and/or reflecting infrared radiation produced by thermal radiation from the Earth's surface.

Greenhouse Gases Observing Satellite (GOSAT). The first Earth observation satellite specifically dedicated to greenhouse-gas monitoring. GOSAT, which was developed by the Japan Aerospace Exploration Agency and launched on January 23, 2009, measures greenhouse gases from 56,000 locations in the atmosphere. The gathered data are shared with NASA and other international scientific organizations.

Greenhouse Gas Industrial Reporting and Control Act. The Greenhouse Gas Industrial Reporting and Control Act enables the BC government to set performance standards for industrial facilities or sectors in order to uphold the Province's commitments to having the cleanest liquefied natural gas (LNG) operations in the world.

Industrial Ecology. Industrial Ecology refers to the practice and study of material and energy flows through industrial systems. Ideally, under the practice of industrial ecology, all waste streams would become valued inputs to other industrial process.

International Emissions Trading Association (IETA). IETA is a non-profit business organization established in 1999 to serve businesses engaged in emerging carbon markets. With its head office in Geneva, Switzerland, IETA's stated objective is to "build international policy and market frameworks for reducing greenhouse gases at lowest cost." The association provides advocacy support for over 150 member companies, in terms of government policy, and is a regular and well-profiled participant at the annual Conference of the Parties.

International Pacific Salmon Fisheries Commission (IPSFC). The IPSFC was a regulatory body run jointly by Canada and the US. The commission's mandate was to protect stocks of the five species of Pacific salmon. The IPSFC operated from 1937 until 1985, at which time the Pacific Salmon Commission was established to enforce the Pacific Salmon Treaty, ratified by Canada and the US in 1985.

Joint Implementation (JI). Joint Implementation is defined in Article 6 of the Kyoto Protocol as a mechanism that allows a country with an emission reduction or limitation commitment under the Kyoto Protocol (Annex B Party) to earn emission reduction units (ERUs) from an emission reduction or emission removal project in another Annex B Party.

Koa Tree. The *Acacia koa* is a species of flowering tree in the pea family, Fabaceae. It is the second most common tree in Hawaiian Islands, where it is endemic. Due to its unique grain and colour, Koa wood is popular for use in furniture making and artwork.

Kyoto Protocol. The Kyoto Protocol is an international treaty which extends the 1992 United Nations Framework Convention on Climate Change (UNFCCC) committing State Parties to reduce greenhouse gases emissions. The Kyoto Protocol, which established targets and timetables for participating parties, was adopted at COP 3 in Kyoto, Japan, on 11 December 1997 and entered into force on 16 February 2005. There are currently 192 parties to the protocol. Canada withdrew from the protocol in December 2012.

Low Energy Nuclear Reactions (LENR). A class of reactions initially referred to as "cold fusion" when introduced by Professors Pons and Fleischman working at the University of Utah in 1989. LENR has not been embraced by the broader scientific community, but research continues with such organizations as Stanford Research Institute International in California, supported in part by Brillouin Energy, and CleanPlanet operating at Tohoku University in Japan, with the participation of the Japanese government, Toyota, Nissan and Mitsubishi Heavy Industries.

Methane (CH_4). Methane is an important greenhouse gas as there are many opportunities in waste management and agriculture to achieve cost-effective CH_4 reductions. Methane has a global warming potential of 21 to 38 times that of CO_2.

Millennium Ecosystem Assessment. The assessment, portrayed to be the first comprehensive audit of the status of Earth's natural capital, was launched in 2001 by the UN and undertook to determine the status of the planet's ecosystems. The assessment grouped benefits into: (a) provisioning services (e.g., the production of food and clean water); (b) regulating services (e.g., climate regulation); (c) supporting services (e.g., the production of nutrients and pollination); and (d) the provision of cultural benefits. The assessment reported that 60 per cent of the 24 key ecosystem services examined were being degraded.

Montreal Protocol on Substances that Deplete the Ozone Layer. An international treaty designed to protect the ozone layer of the atmosphere by phasing out the production of some substances considered to be responsible for ozone depletion. The protocol was agreed to in 1987 and entered into force on August 26, 1989. Current projections indicate that the ozone layer will return to 1980 levels between 2050 and 2070, indicating the protocol's success.

Natural Capital. The planet's stock of natural resources, which includes geological resources, soils, air, water and all living organisms. Natural capital provides humanity with a very wide range of tangible assets and "ecosystem services" that make up the foundation for the world's economy.

Nitrous Oxide (N_2O). N_2O is a significant greenhouse gas, which if considered over a 100-year period, is calculated to have between 265 and 310 times the global warming potential (GWP) of CO_2.

Non-Governmental Organization (NGO). A voluntary not-for-profit citizens' group of members which is organized on a local, national or international level.

North American Space Administration (NASA). NASA is an independent agency of the executive branch of the US federal government. NASA is responsible for the civilian space program as well as aeronautics and aerospace research. The agency began operations in October 1958 and in so doing replaced the National Advisory Committee for Aeronautics.

Ocean Carbon Science Inc. (OCS). OCS was a research collaborative established in 1999 to study carbon, energy and micronutrient flows and dynamics in ocean systems. OCS was intended to support and coordinate research pertaining to remote sensing from satellites and the measurement of carbon sequestration and flows in oceans. In taking on this role, OCS became the lead industry partner in a broader collaboration, supported by the Natural Sciences and Engineering Research Council of Canada and the University of Hawaii, the University of British Columbia and Dalhousie University in Halifax. A lapse in industry support and participation resulted in the company ceasing operations in 2002.

OCO-2. OCO-2 is an American environmental science satellite launched on July 2, 2014 by NASA to replace the Orbiting Carbon Observatory, which was lost in a 2009 launch failure. OCO-2 is a key component of a proposed global fleet of monitoring satellites directed to corroborate emission reporting from the 196 UNFCCC member states.

Ogonis. The Ogonis are a people that number about 1.5 million and live in a 1,050-square-kilometre homeland in southeast Nigeria which they also refer to as Ogoni, or Ogoniland. Also known as the Ogoni Kingdom, the Ogonis drew international attention to resource development on and adjacent their lands through a public protest directed to Shell Oil and the subsequent execution of Ken Saro-Wiwa, one of the Ogoni leaders.

Ohi'a Tree. The most common tree in Hawaii, the Ohi'a, also known as Lehua, is a flowering species of evergreen in the myrtle family. The Ohi'a tree, which is endemic to Hawaii, has flowers showing a range in colour from a brilliant red to yellow. Hawaiian traditions hold that the Ohi'a forests are the home of the goddess Laka. The Ohi'a has been impacted by historical unsustainable harvest and by a host of invasive species.

Pacific Carbon Trust (PCT). The Pacific Carbon Trust was established by the BC government in 2008 to help develop a carbon-offset business sector and to secure carbon offsets to address the government's carbon-neutral operations objectives. Following criticism from the auditor general, as well as hospitals and schools, the PCT ended operations in 2013. Subsequently the Carbon Investment Branch was established to secure a supply of carbon offsets to meet the government's carbon-neutral operations policy and regulations.

Paris Agreement. The Paris Agreement was developed at COP 21 held in Paris in December 2015. The agreement was developed within the framework of the UNFCCC dealing with greenhouse gases emissions mitigation, adaptation and finance starting in the year 2020. The agreement requires ratification by 55 UNFCCC Parties, accounting for 55 per cent of global greenhouse gas emissions, before it comes into force. Article 5.2 of the Paris Agreement recognizes the importance of "non-carbon benefits."

Photosynthesis. The process by which plants, some bacteria and some protistans harness the energy from sun to remove CO_2 from the atmosphere and produce sugar, which cellular respiration converts into

ATP, the "fuel" used by virtually all life on earth. Photosynthesis is currently the only practical method for removing large volumes of CO_2 from the atmosphere.

Phytoplankton. The microscopic photosynthesizing organisms that occur in the upper layers of almost all oceans and bodies of fresh water. Phytoplankton are agents for "primary production," i.e., the creation of organic compounds from carbon dioxide dissolved in the water. Phytoplankton account for approximately half of all photosynthetic activity on Earth and form the basis of the complex ocean ecosystems. They are consumed by zooplankton, small fish, filter feeding invertebrates and whales.

Reducing Emissions from Deforestation and Forest Degradation (REDD). REDD is a mechanism that has been under development by the United Nations Framework Convention on Climate Change (UNFCCC) since it arose as an initiative at COP 11 in 2005. The objective of REDD is to mitigate climate change through reducing net emissions of greenhouse gases by means of enhanced forest management in developing countries.

Reducing Emissions from Deforestation and Forest Degradation + (REDD+). The REDD+ mechanism arose from COP 13 in 2007 and was intended to add the conservation and the sustainable management of forests and the enhancement of forest carbon stocks to qualifying climate actions falling under REDD.

School of Ocean and Earth Science and Technology (SOEST). The School of Ocean and Earth Science and Technology, located at the University of Hawaii at Mānoa, on the Island of Oahu.

SCIAMACHY. The scanning imaging absorption spectrometer for atmospheric chartography was a spectrometer designed to measure sunlight scattered by the Earth's atmosphere or surface in the ultraviolet, visible and near infrared wavelength regions. The instrument was aboard Envisat and became inoperable with Envisat's loss of contact with ground control.

Specified Gas Emitters Regulation (SGER). Alberta was among the first jurisdictions to promulgate a greenhouse gas regulation. Established in 2007, the Specified Gas Emitters Regulation required facilities that emit more than 100,000 tonnes of greenhouse gases a year to reduce their emissions intensity by 12 per cent below their 2003–2005 emissions average. An emitting facility could either achieve the reduction within its fenceline, acquire acceptable offsets, or pay a fee of $15 per tonne of CO_2 to comply with the regulation. A Ministerial Order in 2015 amended the fee to $20 in 2016 and $30 in 2017.

Spectrometer. An apparatus designed to measure the intensity of a spectrum of wavelengths. The spectrometer may be used to graph intensity as a function of wavelength, frequency, energy, momentum or mass. An integrated plant fluorometer is a specialized spectrometer designed to measure the wavelength emitted from chlorophyll, i.e., near red and infrared.

Stanford Research Institute (SRI) International. SRI International is an American non-profit research institute headquartered in Menlo Park, California. The trustees of Stanford University established SRI in 1946 as a center of innovation to support economic development in the region. SRI's mission is to create world-changing solutions to make people safer, healthier and more productive. SRI has a long history in the investigation of LENR phenomena.

Sustainable Development Council (SDC). Established by Westcoast Energy Inc. in 1993, the SDC was an integrated sustainable development policy think tank and governance reporting structure. The SDC was comprised of senior managers and executives from 12 companies operating in the natural gas industry across Canada. The council was supported by the Sustainable Development Office, a secretariat staffed with professionals holding capacity in economic analysis, environmental law and emissions trading and management. The SDC was critical to incubating the GEMCo initiative in 1996.

Thermal Radiation. The phenomenon by which all matter emits electromagnetic radiation in direct correlation with its surface temperature measured in degrees Kelvin above absolute zero. This radiation represents the conversion of a body's thermal energy into radiant electromagnetic energy. It is a spontaneous process of the radiative distribution of energy. The quantity of energy radiated per unit surface area is defined by the Stefan–Boltzmann law, while the wavelength (spectrum) of the radiation is defined by Wien's displacement law. The relevance of these laws of physics to climate theory is that greenhouse gases are opaque to much of the thermal radiation being emitted from the Earth and therefore absorb or reflect it, thereby trapping the heat.

TZ1. TZ1 is the original name for what is now known as the Markit Environmental Registry. The acronym refers to the first Time Zone, which is located east of the International Date Line and includes New Zealand.

United Nations Framework Convention on Climate Change (UNFCCC). The United Nations Framework Convention on Climate Change (UNFCCC) is an international treaty that was negotiated at the Earth Summit in Rio de Janeiro in June 1992. The overarching UNFCCC objective is to stabilize greenhouse gas concentrations in the atmosphere at a level that would prevent dangerous anthropogenic interference with the climate system. The UNFCCC entered into force March 21, 1994.

Voluntary Carbon Market. The voluntary carbon market functions outside of the compliance markets. Voluntary markets allow companies and individuals to purchase carbon offsets on a voluntary basis, in order to meet corporate sustainability objectives or brand their operations and/or products and services carbon neutral, climate friendly or some other variant on the theme of lowered carbon emissions. The value of offsets traded in 2015 was US$278-million.

Western Climate Initiative (WCI). The Western Climate Initiative was started in February 2007 by the governors of the western US states of Arizona, California, New Mexico, Oregon and Washington, with the goal of developing a multi-sector, market-based program to reduce greenhouse gas emissions,. Since then, membership has grown and waned. By 2017, WCI activity included the execution of an agreement between Quebec and California to reduce carbon-based emissions and other greenhouse gases through a linked cap and trade system. The agreement covers large emitters, including cement plants, aluminum plants and transportation fuel producers.

Zooplankton. Microscopic animal or animal-like heterotrophic organisms occurring in oceans, lakes, rivers and other fresh water bodies. Zooplankton are comprised of freely floating protozoa, small crustaceans and the eggs and larvae of fish and invertebrates. Zooplankton graze on phytoplankton and other microorganisms, forming an integral part of the oceans' food chain. Zooplankton are consumed by a wide range of organisms, including small and very large fish (e.g., the whale shark and basking shark), filter feeding invertebrates and whales.

Notes

1 I.S. Shklovskiĭ and Carl Sagan, *Intelligent Life in the Universe* (Boca Raton, Fla.: Emerson-Adams Press, 1998; first published 1966).

2 "Jacques Cousteau at NASA Headquarters and The Space Colonies Idea 1969–1977," National Aeronautics and Space Administration, last modified July 10, 2002, http://settlement.arc.nasa.gov/CoEvolutionBook/JCOUST.HTML.

3 Debbie Brill was a member of Canada's National Olympic Team. As a high jumper, Debbie won over 65 national and international championships, and was appointed an Officer of the Order of Canada in 1983.

4 Toronto: Random House, 2013.

5 The first guitar in space was a St. Petersburg-built acoustic the Russians had on board *Mir*. Chris Hadfield took an electric SoloEtte to *Mir* on STS-74 in 1995.

6 The bio-available energy is in the form of adenosine triphosphate, or "ATP."

7 The Environment and Land Use Committee, or ELUC, was a BC cabinet committee, and the secretariat served the committee.

8 Salmonids are a family of native fish that includes in BC five species of salmon, two species of trout and two species of char. Most native salmonid species have populations with an "anadromous" life history, which means they live in both fresh and salt water for part of their life histories. The exception is the lake "trout," which is actually a char that only lives in fresh water.

9 Cutthroat trout do not die upon spawning. Nor do rainbow trout, better known as steelhead if they go to sea.

10 Ken Kristian, "The Greatest Salmon River In The World – The Fraser," last modified March 18, 1998, http://www.virtualnorth.com/fireside/thefraser.html.

11 Otherwise known as "poachers."

12 All true salmon die upon spawning, and these fish had stopped eating long ago. The radiotags were occupying empty stomachs and had virtually no impact on the migrating salmon.

13 White sturgeon can weigh over 500 kilograms.

14 Assigned a value.

15 Some of the food and waste challenges were depicted by the botanist astronaut Mark Watney, played by Matt Damon, who was abandoned on Mars for two years in the 2015 film *The Martian.*

16 Other than a brief turn at tracking radiotagged sockeye salmon over the Fraser River in a small Cessna aircraft.

17 Freeman Dyson, *Weapons and Hope* (Toronto: HarperCollins, 1984); Freeman Dyson, *Advanced Quantum Mechanics* (Hong Kong: World Scientific Publishing, 2004).

18 Kenneth Brower, *The Starship and the Canoe* (Toronto: HarperCollins, 1983).

19 The other scientists included Richard Feynman, a theoretical physicist who would solve the riddle of the *Challenger* space shuttle disaster, Robert Oppenheimer, who led the Manhattan Project, and Edward Teller, nicknamed "Father of the H-bomb."

20 Handwritten letter from Freeman Dyson dated August 6, 1983.

21 For example, John Muir: "When we try to pick out anything by itself, we find it hitched to everything else in the Universe"; and Serge Kahili King: "We are all connected to everyone and everything in the Universe."

22 Tom Hawthorn, "Kayak Builder a Prophet from the Wilderness," *The Globe and Mail*, May 22, 2012, http://www.theglobeandmail.com/news/british-columbia/kayak-builder-a-prophet-from-the-wilderness/article4198917/.

23 George Dyson, *Baidarka* (Anchorage: Alaska Northwest Books, 1986).

24 In 1991 I was awarded a Canadian patent for growing wood in a process from cambium re-engineered to have planar rather than cylindrical geometry, which accommodated continuous non-destructive harvest of designed wood products.

25 These circumstances related to departmental politics.

26 A tree's cambium can be seen as a cylinder of embryonic tissue that lies between the bark (outside) and the wood (inside). This tissue goes through cell division to produce wood (or xylem) cells on its inside and bark (or phloem) cells on the outside. The "cork" we use to seal wine bottles is the bark tissue of an oak tree.

27 The analyses included measures of lumen diameter and cell wall thickness, which relate to wood quality.

28 From the Moody Blues album *In Search of the Lost Chord* (1968).

29 Current atmospheric CO_2 concentrations are in the 400+ ppm zone, and concentrations are rising by approximately 2 ppm annually.

30 The thesis, titled "Cambial and Photosynthetic Activity Relations in Untreated, Wounded, and Geotropically Stressed White Spruce (*Picea Glauca* [Moench.] Voss) Seedlings," was defended on February 28, 1990. The external examiner was Professor John Gordon, Dean of the Yale School of Forestry.

31 Trees in cross-section are generally round. Trees grown under the constant influence of wind can grow their stems in a manner that buttresses them against the wind forces, by laying more wood down on the windward and leeward sides of the stem.

32 Work on eccentric tree growth under wind conditions continues. E.g., see "Measurement of Pine Tree Sway; Implications for Wind-caused Eccentricity and Damage," Association of American Geographers, Annual Meeting, Seattle, April 10–16, 2011.

33 Dr. Krajina was invested into the Order of Canada in 1981 for his outstanding contributions to the science of Ecology.

34 Patricia Davies, "Diversions," *The Globe and Mail Report on Business Magazine*, October 1989, 19.

35 "If You Go Down to the Woods...," *The Economist*, April 15, 1989, https://www.highbeam.com/doc/1G1-7520077.html.

36 Michael Crichton, *Jurassic Park* (New York: Alfred A. Knopf, 1990).

37 Patrick Moore, *Green Spirit: Trees Are the Answer* (Toronto: Hushion House, 2000).

38 I was asked to help with the selection of suitable directors for the foundation. An amazing orator, David attracted overflow crowds to his genetics lessons at UBC when I was an undergrad there. Atypically for a professor, David could often be seen playing Frisbee and joking with students in the UBC Biosciences Building courtyard.

39 The other main greenhouse gases (CH_4, HFCS, N_2O, etc.) are always converted to CO_2 equivalents, or CO_2e, based on their radiative forcing potentials. For example natural gas, which is virtually pure CH_4, has 21 times the radiative forcing of an equivalent measure of CO_2.

40 Now Spectra Energy, after having been acquired in 2002 by Duke Energy, headquartered in the US.

41 I would present reports on all environmental matters to the Environment, Health, and Safety Committee of the board of directors.

42 I would become the first chairman of CEPA's Greenhouse Emissions Management Committee.

43 The CGA's Environment Committee held quarterly meetings across the country.

44 At this point I reported to Westcoast's CEO Mike Phelps that "climate change" would become an election issue. I had no idea it would be 2015 before the issue would truly arrive, or that there would be an associated cabinet position, i.e., the Minister of Environment and Climate Change, appointed by Prime Minister Justin Trudeau in November 2015.

45 Personal communication, email dated June 2011.

46 Canadian Gas Association, Canadian Energy Pipeline Association, and Canadian Association of Petroleum Producers.

47 The Red Book contained the Liberal Party's green policy platform.

48 In this case "they" being energy industry players.

49 Natasha Mekhail, "Godfather of Green," *Alberta Venture*, August 1, 2008, http://albertaventure.com/2008/08/godfather-of-green/.

50 In the parlance of carbon trading the term used is "fungible."

51 There are a number of protocols/methodologies that exist for different project types and emission reduction programs: the Gold Standard; ISO 14064-2; VCS (Verified Carbon Standard); and a number of jurisdictional protocols such as California's ARB Compliance Offset Protocols.

52 Kelley Hamrick and Allie Goldstein, *Raising Ambition: State of the Voluntary Carbon Markets 2016*, ed. Molly Peters-Stanley and Gloria Gonzalez, Ecosystem Marketplace: A Forest Trends Initia-

tive, 2016, http://forest-trends.org/releases/uploads/SOVCM2015_FullReport.pdf.

53 In a cap and trade system, a maximum amount of emissions (the cap) is set for a defined group of emissions sources and emitters. Serialized units in this amount (called allowances) are issued into the market. Emitters covered by the regulations have to submit allowances equal to their emissions. The market sets the price of carbon based on supply and demand of the allowances. Such schemes can also recognize carbon offset credits from emission reductions that occur outside the geographic and emission sources boundary set for the allowances. They may also recognize allowances from other jurisdictions' cap and trade schemes.

54 This objective may be regulated by government or self-imposed by a voluntary corporate environmental policy. In the case of the latter, we are referring to the "voluntary carbon offset market." This voluntary market, which began almost 30 years ago, is where most of the early learnings around offsetting have occurred.

55 Costs of validating methodologies and certifying reductions or removals can and have driven projects into a net loss position. Hundreds of thousands of dollars or more can be spent on validation and certification processes.

56 Perry Toms is currently CEO of Steeper Energy, Calgary, Alberta.

57 Murray represented New Zealand in the negotiations at the 1997 COP in Kyoto, Japan, where emissions trading was first heralded, and chaired the technical process that then produced the detailed rules for international emissions trading and project-based trading between developed countries adopted for the Kyoto Protocol.

58 As I was not representing any of the GEMCo member companies, I immediately resigned from GEMCo and flew south by southwest to contemplate my next chapter under the banyan trees in Lahaina.

59 Methane is 21 to 38 times more potent a greenhouse gas than CO_2.

60 There have been suggestions that the number is actually much higher due to unreported coal power plant emissions.

61 J.T. Houghton et al., eds., *Climate Change 2001: The Scientific Basis* (Cambridge: Cambridge University Press, 2001).

62 Monika Rhein et al., "Observations: Ocean," in *Climate Change 2013: The Physical Science Basis* (Cambridge: Cambridge University Press, 2013), https://www.ipcc.ch/pdf/assessment-report/ar5/wg1/WG1AR5_Chapter03_FINAL.pdf.

63 Ibid.

64 Temperatures also impact the flow of CO_2 between the atmosphere and ocean, with higher temperatures releasing more CO_2 from the ocean.

65 "High Nutrient Low Chlorophyll Ecosystems," School of Ocean and Earth Science and Technology at University of Hawaii, http://www.soest.hawaii.edu/oceanography/courses/OCN626/2008_OCN%20626/HNLC%20regions%20lecture_2008.pdf.

66 Paul Preuss, "Asian Dust Storm Causes Plankton to Bloom in the North Pacific," Berkeley Lab Research News, last modified October 24, 2002, http://www2.lbl.gov/Science-Articles/Archive/ESD-Gobi-plankton-Bishop.html.

67 Charles Graeber, "Dumping Iron," *Wired*, November 1, 2000, http://www.wired.com/2000/11/ecohacking/.

68 SOFeX is an acronym for Southern Ocean Iron Experiment, playing on the chemical symbol for iron, *Fe*.

69 John Martin's titillating quote was: "Give me half a tanker of iron, and I'll give you the next ice age."

70 Preuss, "Asian Dust Storm," http://www2.lbl.gov/Science-Articles/Archive/ESD-Gobi-plankton-Bishop.html.

71 Ibid.

72 See UN Environmental Programme, *Global Deserts Outlook*, http://www.unep.org/geo/gdoutlook/046.asp.

73 This query was an echo of Jacques Cousteau's comments back in 1978 (see Chapter 1).

74 I have always ensured that any carbon offset programming I was involved in would bring multiple benefits, i.e., ecological restoration, biodiversity and/or community benefits.

75 Some months later, the NGO acknowledged having erred in its calculations of the volume of residual hydrocarbons.

76 The three major universities were: the University of British Columbia (Vancouver), Dalhousie University (Halifax) and the University of Hawaii (at Mānoa).

77 ASLO had hosted the ocean carbon science workshop in Washington, DC, in April 2001.

78 Sure enough, seeing no light at the end of the tunnel, our industry partners began to withdraw their support a few months later.

79 Remarkably, several days earlier, when for health reasons an Edmonton restaurant refused to prepare Carlton a hamburger cooked "rare," he reportedly commented, "Sometimes I think we worry about the wrong things."

80 Quirin Schiermeier, "Climate Change: The Oresmen." *Nature* 421, no. 6919 (2003): 109–10, http://dx.doi.org/10.1038/421109a.

81 Joe Castaldo, "Something in the Water," *Canadian Business*, accessed January 9, 2017, http://site.canadianbusiness.com/longform/something-in-the-water/.

82 "Iron Fertilization," *Living on Earth*, accessed January 9, 2017, http://loe.org/series/series.html?seriesID=27.

83 "Salmon and Ocean Iron Fertilization," *Eating Jellyfish: Or Can We Protect and Recover Marine Ecosystems?*, last modified December 8, 2012, http://eatingjellyfish.com/?tag=haida-gwaii-and-iron-fertilization.

84 "Two Years After...." *Planet Experts*, last modified September 2, 2014, http://www.planetexperts.com/two-years-russ-george-illegally-dumped-iron-pacific-salmon-catches-400/.

85 For example, Fraser River sockeye returns in 2010 were an estimated 34 million fish. The returns for the year previous were 1.7 million fish.

86 EPRI had, however, stopped its work in LENR following a lab explosion that took the life of one of its scientists.

87 An unusual first name, but it turned out each of his siblings was named after a sailboat part.

88 Branson's interest in carbon and climate change was confirmed in Cancun at the UNFCCC's COP 16 in 2010, where ERA's Erin Kendall (white cowboy boots) crossed paths with him on several occasions as he darted incognito among events.

89 Western Wind Energy became Canada's first publicly trading wind energy company in 2001, and immediately began developing project opportunities in Arizona and California.

90 Coincidentally, just hours before Sir Arthur's passing on March 19, 2008, a massive gamma ray pulse, known as GRB080319B, hit planet Earth, setting a new record as the farthest object that could be seen from Earth with the unaided eye. It was suggested by science writer Larry Sessions that the burst be named "The Clarke Event."

91 This took a great deal of explaining, as the carbon world was poorly understood (and still is).

92 The "Millennium Ecosystem Assessment" reported that ecosystems were being degraded at a global scale. Our presentation to the mayor of Maple Ridge suggested that our CERP programming in Maple Ridge would have a ripple effect and be emulated in other jurisdictions. Over the next several years several municipalities and cities engaged in CERP programming.

93 BC Hydro did not buy offsets but graciously offered a modest contribution to support the Community Ecosystem Restoration

Project's kickoff. To date, CERP has registered 1.6 million tonnes of carbon offsets to supply the voluntary carbon market.

94 Bart always had his wheeled cart, with his PC and coveted silviculturist's Red Book, presumably so that he could address any tricky technical questions regarding restoration prescriptions, carbon sequestration rates etc.

95 Over several months, we would raise approximately $4.5-million in equity investments, with a company valuation of $17-million.

96 ERA continued to merge and acquire suitable companies as its business model evolved to address changing markets. The company has now expanded its focus beyond carbon and into agro forestry. The company was been rebranded in 2014 and is now NatureBank. Naturebank trades on the TSX Venture Exchange under the symbol COO.V.

97 The criteria of the Chicago Climate Exchange (CCX), established in 2003 by Richard Sandor, were not workable for ERA's Community Ecosystem-based offsets.

98 HEAG Sudhessische Energie AG (HSE) is a German-based energy and infrastructure company based in Darmstadt. Together with its subsidiaries, HSE provides retail, industry and residential communities with electricity, natural gas, water and technical services.

99 On May 31, 2010, ERA announced the sale of $800,000 worth of "Verified Emission Reductions" with HSE.

100 As one of Germany's largest electrical and natural gas suppliers, Entega Vertrieb GmbH & Co. KG (Entega) is focused on the provision of modern, sustainable energy.

101 Or so it appeared.

102 For the story of Markit, see Chapter 12

103 The Gulf Island project won the Premier's Award for Innovation in BC in 2010–11.

104 See http://www.ucsusa.org/global_warming/solutions/stop-deforestation/deforestation-global-warming-carbon-emissions.html#.WKxhFPkrK71.

105 Forest Trends' REDDX Initiative reports that the private sector accounted for 10 per cent of finance tracked through 2014.

106 TZ1 denoted "Time Zone One," as the organization was established in New Zealand.

107 The TZ1 Registry (part of TZ1 Market) was founded by the New Zealand stock exchange (NZX) in 2007 and sold to Markit Group Limited, where it is known today as the Markit Environmental Registry.

108 "Overview of the Millennium Ecosystem Assessment," *Millennium Ecosystem Assessment*, accessed January 9, 2017, http://www.millenniumassessment.org/en/About.html.

109 "What Are our Closest Animal Relatives?" Science Museum, accessed January 9, 2017, http://www.sciencemuseum.org.uk

/WhoAmI/FindOutMore/Yourgenes/Wheredidwecomefrom/Whatareourclosestanimalrelatives.

110 Government data have shown that over 50,000 orangutans have already died as a result of deforestation due to palm oil cultivation in the last two decades. See http://www.saynotopalmoil.com/Whats_the_issue.php.

111 For example, Dole Pineapple got its start on the Island of Lanai in the early 1920s but has long since left due to competition.

112 The Parker Ranch on the Big Island was once the largest ranch in the United States. Operations have virtually ceased, owing to the cost of shipping live cattle to mainland slaughterhouses.

113 Scientists generally do not use such words, but it is the only word that captures this experience.

114 Plants use this CO_2 and sunlight to synthesize food and building blocks for plant tissue, such as wood.

115 Fluorescence only occurs in light, when plants are photosynthetically active. If this were not the case, forests, crops and ocean plankton would be visibly glowing red at night. Fluorometers are able to filter out non-fluorescent wavelengths in daylight, in order to "see" and measure the red glow. To the unfiltered human eye, the red glow is too dim to see given the very much more intense green reflectance.

116 The wavelengths of this glow are generally found between 650 and 800 nanometres (nm) with two peaks in the visible (685 nm) and near-infrared (740 nm).

117 Real life application of space-borne systems to detect and measure fluorescence in the oceans was seen in Chapter 7, "Carbon and the Oceans."

118 Proceeding in Envisat Symposium 2007, Montreux, Switzerland, April 23–27, 2007.

119 J. Joiner, Y. Yoshida, A.P. Vasilkov, Y. Yoshida, L.A. Corp, and E.M. Middleton. "First Observations of Global and Seasonal Terrestrial Chlorophyll Fluorescence from Space," *Biogeosciences* 8, no. 3 (2011): 637–51, accessed January 9, 2017, http://dx.doi.org/10.5194/bg-8-637-2011.

120 "New Satellite to Measure Plant Health," European Space Agency, last modified November 19, 2015, http://www.esa.int/Our_Activities/Observing_the_Earth/New_satellite_to_measure_plant_health.

121 "UK Wins Satellite Contract to 'Weigh' Earth's Forests," *BBC News*, May 3, 2016, http://www.bbc.com/news/science-environment-36195562.

122 Chris Corrigan, "Place Names in Howe Sound," *Chris Corrigan: Consulting in organizational and community development*, last modified July 16, 2001, http://www.chriscorrigan.com/miscellany/bijournal/16-07-2001.html.

123 "B.C. Waters Officially Renamed Salish Sea," *CBC News*, July 15, 2010, accessed January 9, 2017, http://www.cbc.ca/news/canada/british-columbia/b-c-waters-officially-renamed-salish-sea-1.909504.

124 As boys are wont to do, and oblivious to any relevant law regarding wildlife, I captured the lizard (this was one of the biting species) and took it home to a terrarium. A few days later, my close friend David Shankie visited the terrarium and queried as to where the "other lizards" had come from. "Lizzie" had given live birth to four fully formed and ravenously hungry babies. Keeping these five lizards fed became quite a chore, as they would only eat live spiders. Lizzie and her family were soon returned to the islands.

125 "Breaching Humpback Whale in Howe Sound," YouTube video, 0:28, posted by Terry Peters, October 3, 2013, accessed January 9, 2017, https://www.youtube.com/watch?v=H3cgm4SbRgI.

126 Much of the historical information contained in this chapter is derived from *Around the Sound: A History of Howe Sound-Whistler*, by Doreen Armitage, and published by Harbour Publishing in 1997. Information on the biological resources of the sound was derived from discussions with the curatorial staff and the Royal British Columbia Museum in Victoria, and local residents of the Britannia area.

127 Andy Lamb and Bernard Hanby, *Marine Life of the Pacific Northwest – A Photographic Encyclopedia of Invertebrates, Seaweeds, and Selected Fishes* (Madeira Park, BC: Harbour Publishing, 2005).

128 Aibing Guo, "China Says It's Going to Use More Coal...," *Bloomberg*, November 7, 2016, https://www.bloomberg.com/news/articles/2016-11-07/china-coal-power-generation-capacity-may-rise-19-in-5-year-plan.

129 Marietta Armanyous, "Curbing Coal Production: Is China Pulling its Weight?," NATO Association of Canada, January 11, 2017.

130 Mike Pinder, "Melancholy Man," from the Moody Blues LP *A Question of Balance* (1970).

131 Core Writing Team, R. K. Pachauri, and A. Reisinger, eds., *Climate Change 2007: Synthesis Report. Contribution of Working Groups I, II and III to the Fourth Assessment Report of the Intergovernmental Panel on Climate Change*, IPCC. Geneva, 2007, accessed January 9, 2017, http://www.ipcc.ch/publications_and_data/ar4/syr/en/contents.html.

132 Simon Rogers and Fiona Harvey, "Global Carbon Emissions Rise Is Far Bigger than Previous Estimates," *The Guardian*, June 21, 2012, accessed January 9, 2017, https://www.theguardian.com/environment/2012/jun/21/global-carbon-emissions-record.

133 *United Nations Framework Convention on Climate Change*, Proposal by the President for the Adoption of the Paris Agreement presented in Conference of the Parties, Paris, France, November 30–December 11, 2015, accessed January 9, 2017, https://unfccc.int/resource/docs/2015/cop21/eng/l09.pdf.

134 "Trump Team Memo Hints at Big Shake-Up of U.S. Energy Policy," *Bloomberg*, December 9, 2016, accessed January 9, 2017, https://www.bloomberg.com/news/articles/2016-12-09/trump-team-s-memo-hints-at-broad-shake-up-of-u-s-energy-policy.

135 Accessed January 9, 2017, https://globaloceanforumdotcom.files.wordpress.com/2013/03/final-roadmap-summary-nov-7-2016.pdf.

Dr. Robert William Falls, Ph.D., R.P.Bio., co-founded ERA Ecosystem Restoration Associates Inc. in 2004 and served as its CEO from 2004 to 2011. Dr. Falls has a strong history in the field of climate change, developing and managing projects in China and domestically for corporate clients. He serves as chairman and a director at ERA Ecosystem Restoration Associates Inc. and ERA Carbon Offsets Ltd., and is a senior adviser to the GLOBE Foundation of Canada. He founded and chaired the Greenhouse Emissions Management Consortium (GEMCo), and the Sustainable Development Council for Canada's largest integrated gas company. Since 2004 he has been an adjunct professor at the University of British Columbia's Institute for Resources, Environment and Sustainability. Robert lives in North Vancouver, British Columbia.